HISTORICAL ETHNOBIOLOGY

HISTORICAL ETHNOBIOLOGY

MARIA FRANCO TRINDADE MEDEIROS

Professor, Department of Botany, National Museum/ Federal University of Rio de Janeiro, Rio de Janeiro, Brazil

ELSEVIER

ACADEMIC PRESS

An imprint of Elsevier

Academic Press is an imprint of Elsevier
125 London Wall, London EC2Y 5AS, United Kingdom
525 B Street, Suite 1650, San Diego, CA 92101, United States
50 Hampshire Street, 5th Floor, Cambridge, MA 02139, United States
The Boulevard, Langford Lane, Kidlington, Oxford OX5 1GB, United Kingdom

Library of Congress Cataloging-in-Publication Data
A catalog record for this book is available from the Library of Congress

British Library Cataloguing-in-Publication Data
A catalogue record for this book is available from the British Library

ISBN: 978-0-12-816245-3

For information on all Academic Press publications visit our website at https://www.elsevier.com/books-and-journals

Publisher: Charlotte Cockle
Acquisitions Editor: Anna Valutkevich
Editorial Project Manager: Devlin Person
Production Project Manager: Maria Bernard
Cover Designer: Miles Hitchen

Typeset by TNQ Technologies

Contents

Acknowledgments

Along the trajectory of development of the body of the work, I had the pleasure of having the delicacy and attention of Devlin Person, who always took care of the entire editorial process. I would also like to thank Maria Bernadette Vidhya for her constant attention towards me during the finalization of this edition.

For the use of the images, I sincerely thank the *Pinacoteca de São Paulo* (São Paulo, Brazil) and the library, *Biblioteca José Antônio Gosalves de Mello do Instituto Ricardo Brennand* (Recife, Pernambuco, Brazil). Likewise, I am grateful for the zeal with which Kavitha Balasundaram took care of the images inserted in the book.

As the English language is not my original language, I resorted to the inestimable help of my mother and friend, Maria José Franco Trindade Medeiros, who reviewed and translated the material produced.

I had the joy of doing teaching experiences between 2008 and 2020 with graduate program students in botany, ecology, and ethnobiology and nature conservation of Rural Federal University of Pernambuco (Recife, Pernambuco, Brazil); natural science and biotechnology of Federal University of Campina Grande (Cuité, Paraíba, Brazil); and biological sciences (botany) of Federal University of Rio de Janeiro (Rio de Janeiro, Rio de Janeiro, Brazil). Thus, my involvement in these teaching activities provided experiences that gave me the desire to write a book from the perspective of ethnobiology.

I would not finish my acknowledgments without expressing my gratitude for the stimulus received from colleagues, as well as the necessary environment and conditions I had to write each chapter, either in the Department of Botany, in the Biological Sciences (Botany) Graduate Program, and also in the library of the National Museum of the Federal University of Rio de Janeiro (Rio de Janeiro, Rio de Janeiro, Brazil), which has been my new scientific house since 2018.

CHAPTER ONE

Introduction

Contents

1. Initial considerations

Ethnobiology comprehends, among other things, the study of classification systems of the living world by any culture (Posey, 1987). Or rather, it is the study of the knowledge and conceptualizations developed by any culture to double living beings and biologic phenomena.

The prefix "ethno-" indicates attempts to understand a particular theme (e.g., plants), having as reference the knowledge, beliefs, and practices that a particular social group presents in relation to this theme, as well as their possible connections with the formal knowledge.

The disciplinary clipping in the ethno-scientific field is *sui generis* because the different approaches—ethnobotany, ethnozoology, ethnopedology, ethnomicology, etc.—are not necessarily mutually exclusive.

Consider the following: 1. the variety of approaches, objectives, and methods observed in ethnobiologic studies and 2. the perspective of the articulation of local knowledge with the academic and of the natural sciences

Historical Ethnobiology
ISBN: 978-0-12-816245-3
https://doi.org/10.1016/B978-0-12-816245-3.00001-9

with the social and human sciences; it thus seems adequate to characterize ethnobiology as a field of knowledge crossing (Marques, 2001).

It is very common to associate ethnobiology with the study of indigenous societies. This was a historical limitation imposed by the first ethnographic and anthropologic reports. However, the amplitude of the field allows that if using an adequate methodology, several other approaches can be realized.

The history of ethnobiology can be divided into the preclassical, classical, and postclassical periods, as proposed by Clément (1998). Thus, we will adopt here this classification, punctuating for each period the main authors and events that contributed to the structuring of this scientific discipline.

Within ethnobiology arises ethnobotany, which through its historical and conceptual development helps us shed light on the history and development of ethnobiology. This is due to the fact that ethnobotany is, within ethnobiology, the field of greatest scientific expressivity.

So, we will develop the following: 1. some historical aspects of ethnobiology, because we understand that this broader historic view will lead us to a comprehension of the history and development of the historical development; and 2. history and development of ethnobotany through the evolution of the concepts given to this science. By this analysis, we will arrive at the contributions that considered the material and immaterial goods preserved as a source of research. The proposal is that we conclude with the visualization of the profile of historical ethnobiology as a somewhat recent scientific approach.

2. History and development of ethnobiology

During an expressive period of its history, researches in ethnobiology were realized from information collected by researchers who were dedicated to the understanding of how human groups classified as "primitive" utilized their natural resources (vegetables, animals, etc.) (Castetter, 1944).

Throughout its scientific development, ethnobiology was developed from two main approaches that were adapted to this investigation of "primitive" peoples. One of these approaches is denominated cognitive. In this investigative perspective the concern was to develop studies that sought to register the perception and knowledge of certain cultures about their own existing natural environment (Hunn, 2007). According to this thought, Brent Berlin and William Balée were those who initiated works from this perspective. The second approach is economic. This other perspective goes in a way to study the process in which througth nature's resouces as raw material, these cultures make biocultural artifacts, insumes and other useful products (Hunn, 2007). Ghillean Prance, Elaine Elisabetsky, Claude Lévi-Strauss, Margareth Emmerich and also Brent Berlin, were some of the researchers who initially developed research in this line.

According to Daniel Clément (1998), the history and development of ethnobiology encompasses three phases. But for Eugene Hunn (2007), it would be four periods, and Steve Wolverton (2013) still presents a fifth phase for ethnobiology. Let us see in sequence a description that will jointly present the proposals of these authors for the historical analysis of the development of all these phases of ethnobiology.

2.1 Preclassical period (XIX-1950), Clément (1998)/First phase, Hunn (2007)

Main ideas

Daniel Clément (1998) and Eugene Hunn (2007) present the same characteristics for the preclassical period, or first phase, about the historical evolution of ethnobiology.

For both authors, the preclassical phase of ethnobiology is marked by the following:

- a descriptive approach of plants and animals;
- an essentially ethical perspective, that is, signaled by a posture and academic interpretation in the studies on the behaviors and the relation of the native peoples with the natural resources;
- considered a utilitarian period;
- a focus on registering the popular name of the plants, their use, and performing their botanical identification;
- an interest in the economic utility of products made from plants or animals that could be used by Western society.

The preclassical period of ethnobiology, which occurred at the end of the 19th century, was marked by the realization of researches aimed at understanding and recording the knowledge that different peoples and cultures had about biodiversity. Europeans who dedicated themselves to studies of this nature had an interest in natural resources and their uses by native peoples of the so-called New World, as well as in other regions outside the European continent. The European view on the discovery of this exotic nature was strongly linked to the intention to guarantee for itself the possible economic gains that these resources and the knowledge associated with them could provide them. This phase of the history of ethnobiology will be characterized by a science determined by the model—center (colonizer, Europe) and periphery (colonized, New World and other colonized regions)—that dictated the way in which social relations, economics, and commercial plans, etc., occurred at that time. Following this model, the ethnobiologic research started from the principle that it was necessary to document the use of natural resources, especially of plants and animals, which possessed some lucrative potential for colonizers.

In this perspective of the construction of ethnobiologic scientific knowledge, the naturalists who went through the new territories left an important contribution. It was they who dedicated themselves to carry out a detailed register of the biodiversity, the physical environment, and the native peoples. There was also interest in describing the relation of these local peoples with the flora and fauna of these new environments.

We can say that this way of describing was characteristic, and the scientific making was guided by the utilitarian interest that the colonizing countries had over natural resources of the new known territories. It should be pointed out here that we cannot categorically attest this form of doing science as being carried out by all naturalists and scientists of the preclassic epoch. But, in general, we can affirm that science that was done was not disinterested and realized solely for the sake of the sciences. Despite this aspect that conditioned the researches, this period was fundamental in laying the basis for the development of future studies on natural resources and environments, as well as on the varied ethnicities and their cultures present in the New World (Clément, 1998).

It was in this preclassical period that the first studies of the disciplines that can be considered of great expressiveness were established due to the number of researchers who dedicated themselves to them, and thus, of developed researches. These disciplines are ethnobotany and ethnozoology.

Studies on the interrelations between people and natural resources, particularly plants, as well as the way in which these human populations understood their living environments, with their own biologic phenomena, had a principle of structuring with the works of the North American Harshberger, in the year of 1896 (Clément, 1998), as we will see later in this chapter.

Like Harshberger, other researchers of North American and European origin and with background principally in anthropology were those who acted in a massive way during the preclassical period (Anderson, 2011).

2.2 Classical period (1950–80), Clément (1998)/Second and Third phases, Hunn (2007)

Main ideas

- cognitive and classificatory periods;
- consideration of the cultural context;
- focus on understanding how human groups ordered plants and animals, that is, to establish and classify relations between individuals;
- belief in the existence of cultural domains and concern to know them and see how they are structured to be able to act on the individuals;
- ethnotaxonomy;
- folk taxonomy;
- valuation and incorporation of the emic perspective, that is, the worldview of native peoples themselves.

The classical period, or second and third phases of ethnobiology, had its beginning in the first half of the 20th century and was a phase that became known as cognitive ethnobiology. It is a period in which the most prominent feature was the search for the indigenous knowledge as a mean to understand how people comprehend, interpret, and give meaning to the things that compound their living environment (Clément, 1998). A researcher who stood out in this period and who provided this paradigm of an ethcal science for an emic science and thus provided a movement in the history of ethnobiology was Harold C. Conklin (Hunn, 2007).

Hunn (2007) also states that, in 1954, Conklin conducted research on the nomenclature and botanical classification of the Hanunóo. From this work, Conklin inaugurated the phase of the studies that assume a position dedicated to the detailed research under an emic prism. Thus, the interest of research turned to the local linguistic use in order to well comprehend peoples' life history in their environment.

According to Toledo and Alarcón-Cháires, (2012), Conklin sought to understand in his studies how human beings carry out the process of appropriation of natural resources. This search of Conklin would imply understanding both the biological and physical elements with which people interact. It is worth saying that people interact with plants, animals, abiotic elements and other physical aspects like clime, and so on. Besides that, it was included in his investigative perspective the comprehension about the perceptions and uses that peoples from different cultures make of these resources.

The fact that unifies the works developed in this phase is the great influence of the cognitive approach with the use of cognitive psychology and linguistics to arrive at an understanding about the perceptions of human beings on the elements that compound nature (Hunn, 2007; Albuquerque, 2005). Following this path, other researchers that worked during this period and made important contributions to ethnobiology were also Brent Berlin and William Balée.

Another approach that gained momentum during this time named as classical period by Clément or third phase by Hunn was characterized by the strength of ecology. These studies were intensified especially between the 1970s and 1980s. In this stage of ethnobiology stands out the name of Victor Toledo, a Mexican researcher who proposed a new approach. The effort of Toledo and his collaborators was to provide what until then was considered a fragility in the ethnobiology. In the face of the absence of an amplitude in the scientific interpretation of the ecologic context in which the interrelations take place, they began to be considered by them as a relational system composed by three dimensions: 1. belief systems and people values; 2. people's knowledge about natural resources; and 3. people's management practices of these resources. This scheme was the basis for Toledo to develope and propose a model addressing the interrelations that became known as cosmos-corpus-praxis (or matrix k-c-p) (Hunn, 2007; Toledo and Barrera-Bassols, 2009).

2.3 Postclassical period (after the 1980s), Clément (1998)/ Fourth phase, Hunn (2007)

Main ideas

- Socioecologic studies;
- Establishment of cooperation (scientists and traditional peoples);
- Great concern about the return;
- Principal focus becomes to work with local and academic science (cooperative management of resources);
- Space for intellectual property rights of indigenous peoples discussions;
- Study the system of local organization for the appropriation of common resources;
- To have sensitivity to grasp the rules and sanctions that different communities possess, and especially not forgetting that they base their systems on the worldview;
- Study the history of plant domestication to know how and why it is presented in this way nowadays.

The postclassical period or the fourth phase of ethnobiology comprises a period in which the international scientific community will gather for the first time to discuss their researches and to establish a document that expresses the position of ethnobiologists in relation to indigenous and traditional communities, as well as to bring other guidelines to scientific practice. The great articulator of these events was the anthropologist Darrell Posey. Working in the fields of ethnoentomology and ethnoecology for more than a decade, Posey dedicated himself to study the traditional ecologic knowledge of the *Kayapó* Indians (located in the northern region of Brazil). In 1988, Darrell Posey was at the head of the founding of the International Society of Ethnobiology and organized the first International Congress of Ethnobiology, held in the city of Belém, in the State of Pará, in the northern region of Brazil. In this scientific meeting was elaborated the document that received the name "Declaration of Belém". In this declaration, ethnobiolologists assumed a position in favor of the recognition of indigenous peoples and nonindigenous traditional peoples and their knowledge and practices. It is stated that the respect for this population and their values was the main condition for maintening their good survival (ISE, 2014). The document also brings an emphasis related to the professional practice of ethnobiologists in the consciousness of native populations about their own intellectual patrimony. Another focus is the commitment to provide the population in which the research was developed the return of scientific results in its original language (ISE, 2014). Then, these peoples can be integrated into the research process and also researchers are commited to ethical respect for them.

In view of the events that occurred in this period, named as the fourth phase, Hunn (2007) considers that Posey performed a role of fundamental importance for ethnobiology. It was through this movement that this science began to consider as preponderant the preservation of knowledge of indigenous and traditional nonindigenous peoples. In addition, a great importance was also given to stimulate and preserve the property rights of all knowledge belonging to these peoples or communities. Thus, ethnobiology took on a position that became increasingly sensitive to meet the wants and needs of local communities.

2.4 Fifth phase of ethnobiology, Wolverton (2013)

The fifth phase of ethnobiology comprises the current period in which we are, according to Wolverton (2013). This contemporary phase of the historical development of ethnobiology is characterized by being strongly interdisciplinary in relation to its objects of study. Another remarkable aspect of what is nowadays experienced in ethnobiologic research is the importance of these studies for understanding and evaluating of the context of both environmental changes and transmutations in cultural systems (Wolverton et al., 2014).

Insofar as ethnobiology is established as a field of knowledge crossing (interdisciplinary), the scientific making becomes challenging because the range of possibilities becomes reality before the researcher. This necessity to go beyond the limits imposed by an academic discipline and to take into account the carrying out of researches in collaboration with specialists from the most diverse areas of knowledge is what is required as practice of research for today's ethnobiologists. It becomes an exigence to do an immersion beyond the disciplines of basis of ethnobiology, which are anthropology and biology. What is sought is an increasing dialogue with other areas of research, including applied, such as conservation biology and environmental management and ethics (Wolverton, 2013; Hardison and Bannister, 2011).

In a global analysis, we can say that nowadays there is a strengthening of the ideas that were gradually forming in the previous phases. This process has brought to the present the duty of ethnobiology to be a science that promotes a space favorable to respect of intellectual property rights of indigenous and nonindigenous peoples. One of the current focuses has also been the concern with biocultural conservation, ethics and environmental management that seeks to consider local residents (Wolverton, 2013). So many other themes of interest are presented nowadays. What stands out in all of them is the vocation of ethnobiology for the resolution of environmental and cultural issues at different levels of scale, from local to global, through the "glocal" issues (the term used by Robertson means the fusion of the terms GLObal with loCAL, resulting in the term "GLOCAL") (Robertson, 1992). And, even more, a universal mark is to engage in carrying out an academic practice guided by interdisciplinarity. Observing the reality of COVID-19, we can go beyond and affirm that this aspects already commented will be considered in large scale since all people are being affected in their lives in multiple ways. So, this pandemic experience certainly will bring new manners of being in the world, of how peoples through all the world will apprehend new life situations and transform or not their old conceptualizations and practices. Also, in an academic point of view, we can say that making scientific, ehtnobiological knowledge, is already suffering with this new condition and what is becoming to be in the future is a issue to all of us.

3. Historical development of ethnobotany

One of the objectives of ethnobiology is to associate the knowledge of the natural and human sciences to record the full breadth of knowledge, classification, and use of natural resources from the traditional societies (Posey, 1986). That is, a field that promotes respect for the references of the other, to understand the concepts of cosmology that belong to the group being researched (Posey, 1986).

Then, to develop a work in ethnobiology, it is necessary to ask the following:

- Where do we want to arrive?
- What is the essential question of the work?

After answering these questions, we proceed to choose the appropriate methodology to collect socioeconomic information, inventory the history of the place and the landscape of the people and inventory the traditional knowledge about natural resources (flora, fauna, minerals, etc.). All this, always, occurs with the consent of the group researched.

But to take these steps, we must still have a clear definition of the disciplinary formation that we are following such as botany, zoology, ecology, pedology, etc., especially with the disciplinary field of botany, since this is considered one of the fields of ethnobiology that has concentrated more works. It is the field of ethnobotany. Ideas about the theory and methodology of ethnobotany generated models of belief in how ethnobotanical research should be. Let us now turn to the consideration of the development time of this science through the presentation of the main definitions (concepts) present in the literature for ethnobotanical science that characterize each stage of its own history.

3.1 Evolution of the scientific concepts of ethnobotany

There are two scientific positions in relation to the history and development of ethnobotany: 1. universalist, which encompasses all the history of people's interest in the use that other peoples make of the elements of their natural environment, and 2. particularist, which considers the beginning of ethnobotany, as science, from the emergence of the term ethnobotany and its definition by Harshberger, in 1895–96.

Now, we will make a brief consideration of the development and history of ethnobotany adopting the universalist perspective. Some of the documentary elements that indicate humanity's interest in the use of plant resources by different peoples will be presented, and concepts of ethnobotany spanning 1895 to 96 will be presented that illustrate how this science has developed in conceptual and epistemological terms.

3.1.1 Changes in the interpretation of ethnobotany in the course of history

We recall that Richard E. Schultes pondered that ethnobotany has existed since the beginnings of written history of humanity, and it was only around the last 120 years that its recognition as a scientific disciplinary occurred.

Thus, ethnobotanical study is more recent as a science. Let us see that, before the term being coined and its definition, some researchers will approach what would become the constitution of an ethnobotanical science. This was the case of Augustin P. De Candolle (1819), who defined the study of the relations between people and plants as the field of applied botany. Another researcher was Stephen Powers (1873), who introduced the term aboriginal botany. He used the term aboriginal botany to describe the study of any of the forms of the plant world that Aborigines used, whether for medicine, food, textile, or ornament, for example.

The first proposal of the term "ethno-botany" in the scientific field was pronounced by Harshberger in a lecture conferred in the year of 1895. The publication of the term coined by the researcher was conveyed in the article entitled "The Purposes of Ethno-botany", published the following year, in 1896 (Harshberge, 1896). This was the first proposition of the term, when works with tribes of American Indians about food, shelter, and clothes were carried out. Ethnobotany was the science that studied the plants used by primitive and aboriginal peoples. Harshberger considers that ethnobotany could assist in elucidating the cultural position of tribes using plants for food, shelter, or clothing and that such investigations could clarify the problem of plant distribution in the past. In this way, the term appears for the first time, in 1895, without definition, but pointed out manners in which ethnobotany could serve scientific investigation (Schultes, 1962). The following year, in 1896, Jesse Fewkes published "A Contribution to Ethnobotany" and introduced ethnobotany in the academic field and in the anthropologic literature.

Until the turn of the 20th century, ethnobotanical research consisted of studies aimed at recording the use of native plants, having marked utilitarian bias. A new direction on the way of thinking about ethnobotanical researches was delineated from the works developed by Robbins and his collaborators, in 1916. Under the leadership of Robbins were conducted the first systematic studies of ethnic groups. They defined ethnobotanical science not only as a study of the record of the use of plants, but also of the impressions about the environment revealed through customs and rituals (Castetter, 1944). In synthesis, it would be the study of plants and impressions on the environment.

With this movement of rethinking how to make ethnobotanical science, Robbins and his collaborators reported that ethnobotany was more than collecting plants and providing their popular names, but that it was a scientific work that required scientific methods of investigation. And they suggested that ethnobotanists should explain the deep knowledge about life forms of the plants and their relations from the perspective of the indigenous population. They also stressed that ethnobotanists should ask themselves some questions when conducting their field works: 1. What are the primitive ideas and conceptions about the plants? 2. What are the effects of a particular environment on the lives, customs, religion, thoughts, and daily practice of the occupations of the people studied? 3. What are the uses they make of plants for food, medicine, for the ceremonial purpose, etc.? 4. What is the extent of knowledge of these people about the parts, functions, and activities of plants? 5. In which categories are popular names of plants and words related to plants grouped in the language of the people studied, and what can be learned from the study of these names? (Robbins et al., 1916). In this way, from Robbins and his collaborators, new theoretical and methodological notions began to be introduced to ethnobotany.

A little later, in 1919, Melvin Gilmore emphasized the active modification of the plant world by traditional populations. Initiating a growing interest in traditional management of resources, he considers the history of intervention of the human populations in the environment. Thus, he ends up laying the first bases for the development of historical ethnobotany. In 1932, Gilmore will bring another contribution to the definition of ethnobotany. He defines it as every expression of the traditional knowledge of plants. It would not only be a tribal economic botany, but every extension of traditional knowledge about plants and plant life.

Between 1930 and 1950, structuring begins and a master's program in ethnobotany is established at the University of New Mexico in the United states of America. Edward Castetter was responsible for creating this program (Ford, 1978a), which will be a mark in the history of ethnobotany (and of ethnobiology) by establishing a space for teaching and the regular academic discussion of this science.

During this process of structuring a postgraduate program devoted to ethnobotany, two other researchers have brought important contributions to the thinking and delineation of to what ethnobotany is dedicated. They were Richard Schultes and Volney Jones.

In 1941, Richard Evans Schultes defined the field of ethnobotany as being the study of relations that exist between man and the plant environment (Castetter, 1944). Thus, he extends the possibility of researching with any human group, by leaving aside the inclusion of the term "traditional".

Schultes was one of the pioneers in ethnobotanical studies. He had formation in systematic botany and dedicated himself to researching the flora of tropical America. Working in close contact with the Indians of northwestern Amazonia since 1941, he described the preparation and use of innumerous plants used by these indigenous as hallucinogens and contraceptives, among other applications. In his writings, Schultes emphasizes the importance of the daily conviviality he maintained for many years with the Indians for the work he developed (Schultes, 1962). Through this idea written by Schultes, a guideline was established on how to proceed in the field, a standard of conduct.

In the same year of 1941, Volney Jones published the article entitled "The Nature and Status of ethnobotany". In this work, Jones displaces the focus of ethnobotany and puts it on the interrelation of man—plant/plant—man (Castetter, 1944). With this definition, in the measure that an advance of evolutionary thought is established, which describes the dependence and coexistence of man with plants, there is a return to the idea of "primitive" man. Jones acknowledged that the researchers who preceded him agreed to say that ethnobotany should not be concerned solely with the uses of plants, but with the whole universe of interrelations between primitive man and plants.

As we can observe, since Harshberger the definition of ethnobotany has undergone changes over time, accompanying the formation and evolution of this science (Jain, 1987, 1989; Schultes, 1972; Schultes and Reis, 1995).

Between 1950 and 1970, ethnobotany became more and more equated to linguistic concepts and classification. At the same time, Harold Conklin also highlighted the practice significance of understanding popular classification systems (Conklin, 1954). Meanwhile, the interest in paleoethnobotany increased as archeobotanical techniques emerged and were being improved (Cotton, 1996).

In 1978, Richard Ford brought a new proposal, a *New Synthesis* (proposal of articulations) for ethnobotany (Ford, 1978a). He endorsed that modern ethnobotanists should be able to identify which plants are significant;

discover how people of a specific culture classify, identify, and relate to plants; and examine how the perception of the plant world actually guides their action and concomitantly structures the plant environment. Ford felt it was important to modify Jones's definition (in 1944) to accommodate the changes undergone by this field of research. He concluded that ethnobotany is the study of direct interrelations between men and plants. The suppression of the term "primitive" (primitive people) was to allow the expansion of the field of study. And the addition of the term "direct" allowed the field of research to admit those who were in continuous contact with the plants. In this way, Ford believed that the people who were in direct contact with the plants could classify plants in their own way and generate cultural rules for the manipulation of plant resources and their environment.

The following year, in 1979, Mexican Alfredo Barrera Vásquez aggregated another contribution to the academic understanding about ethnobotany. Vásquez affirmed that the main object of ethnobotany is precisely the study of traditional botanical wisdoms. By including the term "traditional" again, Vásquez brings to discussion the idea of traditionality. What the author promotes is a reflection that will move in the sense of considering that the cultural system is open, that is, changes take place and some elements are excluded while others are incorporated in the way people organize themselves culturally and perpetuate their systems of comprehension and expression not only of the world but also in the world.

In the 1980s the structuring of ethnobotany, and thus of ethnobiology, gained another level in world consideration as a scientific field. A portrait of this maturation was the historical mark of the first edition of the *Journal of Ethnobiology*, in 1981, by the International Society of Ethnobiology (see this publication of the journal in ISE website). Other questions gain strength and guide the scientific making of researches in this area, as can be seen in the definitions that have emerged since the 1980s.

In this sense, Hernández Xolocotzi, in 1983, brought his comprehension about ethnobotany by stating that it would be the scientific field that studies the interrelations that are established between man and plants over time and in different environments. He still adds that the elements of the interrelations between people and plant resources are determined by two factors: (a) the environment (ecologic conditions) and (b) by culture. It is interesting to observe that Xolocotzi will bring, once again, the question of time to be considered in ethnobotanical (and thus ethnobiologic) researches, which is proper of historical ethnobotany (historical ethnobiology). The relations occur in time and are modulated by ecologic and cultural conditions.

Already by 1987, Sudhanshu Kumar Jain disagreed with this understanding for approaching only utilitarian aspects, broadened the concept, covering all aspects of the direct relation of plants and man, whether of a concrete order (such as material use and disuse) or abstract (as symbols of cult) (Jain, 1987).

According to the same author, due to this scope, ethnobotany maintains an interdisciplinary collaboration. This understanding was also confirmed by Elizabetsky (1986), who relates ethnobotany to areas such paleobotany, anthropology, ethnotaxonomy, ethnoecology, ethnopharmacology, ethnoagriculture, and ethnomedicine, among others. Prance (1991) also reinforced the interdisciplinarity with the participations of botanists, anthropologists, ecologists, chemists, forestry engineers, and agronomists in the progress of researches in ethnobotany. The studies of associated disciplines or interdisciplinary studies were already being carried out before the 1980s and were intensified with the contribution of several authors. Some of these contributions may be mentioned, such as those of Bodley (1978), who did the circulation of Peruvian Amazon ethnobotanical data among anthropologists, and Jain (1981), Ford (1986), and Xolocotzi (1987), who also dedicated themselves to researches involving indigenous populations from countries in the Americas that served as the basis for anthropologic works.

Ethnobotanical researches also brought contributions in the area of ecology, such as those developed during the 1980s by Baleé and Gely (1989), Anderson and Posey (1989), and Irvine (1989). These authors evaluated different environmental aspects observed by indigenous communities in the Amazon region or dedicated themselves to the study on the management of palm trees by several populations in South America (Balick, 1984; Clement, 1987).

Coming into the 1990s, there was an increasing systematization of graduate and postgraduate programs in ethnobotany, especially in the United States (Cotton, 1996). In this way, what was observed was a strengthening of the education and formation of the ethnobotanist because of the educational programs that became increasingly available.

In this decade, there was also the development of many projects that focused on the practical application of traditional knowledge of plants and that turned to socioecologic studies, the establishment of cooperation between scientists and indigenous and nonindigenous peoples, and concern for the return and intellectual property rights of local communities (Clément, 1998; Hunn, 2007).

In this context, other forms of understanding ethnobotany have emerged. One of them was presented by the Foundation of Ethnobiology of Oxford, in 1993, which establishes ethnobotany as the evaluative record of knowledge of environmental knowledge that different cultures have accumulated through the millennia. Again, the question of time is considered an important factor as the means by which interrelations occur and knowledge about plants is forged.

In 1994, Gary Martin presented his definition, saying that ethnobotany would be all studies (on plants) that describe the interaction of local populations with the natural environment (Martin, 1995).

One year later, 1995, Janis Alcorn stated it to be the study of the interrelations between human beings and plants inserted in dynamic ecosystems constituted by natural and social components.

Going toward the socioecologic question that has been marking the researchers' perspective on the science to which they dedicate themselves, in 1996, Catherine Cotton maintained this approach and affirmed that ethnobotany encompasses all studies concerning mutual relations between plants and traditional peoples.

In the same year of 1996, the Brazilian Maria Christina de Mello Amorozo took up again the definition of Posey (1986) for ethnobiology and assumed this understanding for ethnobotany. So, for this researcher, ethnobotany would be understood as a discipline that is devoted to the study of the knowledge and conceptualizations that have been developed by human societies regarding the vegetal world. This discipline and research would cover both the way in which a social group classifies and orders the plants, as well as the uses that these people make of these species. Then, Amorozo would turn to the apprehension of how folk classifications occur and to the applicability of plant resources conferred by different peoples.

According to Cardona (1985), this term is found in the literature with two distinct connotations: one that refers to a scientific botany, but which proposes to make a study directed to the comprehension of aspects of use, for example, of a certain ethnicity. In this connotation, ethnobotany would be developed by someone with a background in scientific botany, who would look for articulations and matches between Western scientific classification and local classification. The other connotation would focus on botanical science that has a certain ethnicity. In this way of working, the research would be developed by someone with a background in anthropology. The researcher's interest would be focused on understanding of how a specific ethnicity classifies its natural environment. What Amorozo proposed is that the two strands could find a path of complementarity. In this sense, the researcher pointed to the establishment and strengthening of researches that would promote the cooperation between scientists in the field of natural and human sciences, which are areas that are the basis for an ethnobotanical investigation.

Coming to the 2000s, Gerald Wickens gave his contribution by affirming that ethnobotany encompasses all studies concerning mutual relations between traditional peoples and the present and past use of native plants and primitive cultivars, the latter not necessarily natives. It is interesting to note that he does not isolate peoples by considering the relations that are between them. Besides that, Wickens incorporates the temporal question in the construction of local knowledge and ethnobotanical study.

Another important consideration was given by Ulysses Albuquerque, in 2002, when the Brazilian researcher points to ethnobotany as a study of the direct interrelation between people of the existing cultures and the plants of their environment. He allies cultural and environment factors, as well as the conceptions developed by these cultures on plants and their uses. To Albuquerque, ethnobotany seeks to observe relations or processes established between people and plants. The researcher will deal with temporality by conditioning the studies to the people of cultures still alive. Years later, in 2010, Ulysses Albuquerque together with Julio Hurrell stated that ethnobotany studies the interrelations between people and plants from the point of view of Western science (Albuquerque and Hurrell, 2010).

Closing the explanation about the history of ethnobotanical science, we have the understanding that what guides the investigative look, delineates the research, and what it is produced of knowledge in ethnobotany has correspondence with the paradigms of science practiced in the academy.

Expanding what we have seen so far for the development of ethnobiology, we can perceive that throughout this historical process of deepening, expanding, and consolidating this science, the researches in ethnobotany were those that gained more space when compared to other areas of study inserted in the scope of ethnobiology.

Among the varied forms of approach presented by the researches we have the descriptive form, which aims to describe, as the term says, the relations of people with the plants and their uses. Another approach adopted by the researchers is the quantitative approach, which brought the use of statistical tools, so one can test hypotheses in order to understand what is behind people's choice when selecting the use of certain plants instead of others (e.g. Begossi, 1996; Phillips and Gentry, 1993). There is also the particularly historical approach, the central theme of this book. As we will see in the next chapter, the importance of historical ethnobotany (or ethnobiology) lies in the fact that it sharpens the study of relations between peoples and the plant universe by taking into account a broader context of these interactions. What is analyzed in this approach are the traits that reveal the temporal and dynamic changes in the social, cultural, political, economic, and historical aspects of different cultures (Medeiros, 2009). More recently, ecologic and evolutionary approaches have gained space in ethnobiological investigations, considering them necessary to understand the factors that shaped behaviors and forged the knowledge about the elements of the nature present today in the most varied cultures considering their modulation along the history (Albuquerque, 2013; Albuquerque and Medeiros, 2013). Thus, different approaches have sought as well as new ones always seek to expand scientific knowledge about the interrelations between people and plants.

After going through the path in which the scientists dedicated to ethnobotany have worked on conceptualizing the term for over a century, let's move on to another form of historiographic analysis. Now, the perspective adopted will be to draw an arc of dialogue between these definitions and the broader historical contextualization that dates back to the period when people began the tradition of recording local wisdom and practices on the use of resources present in their living environments, until the present moment. It is important to emphasize once again that this history is closely linked to the development of ethnobiology and is therefore indispensable for its deep understanding.

3.2 Brief comment about the history of the development of ethnobotany

3.2.1 *Botany, ethnobotany, and explorers in the New World*

Although people domesticated plants very early in human history, a systematic study of plant resources or nature did not occur quickly. There are occasional plant drawings by people who lived during the Paleolithic period, but they are not as common as animal drawings (Baker, 1968).

Historically, we may situate the interest in the field of ethnobotany as belonging to the explorers and adventurers of Europe. These were the first to observe and document the uses of plants by indigenous peoples that they encountered and met during their expeditions. These may be considered the first discoveries about botany in the so-called New World. This discovery of the botanical universe so exotic to foreign eyes marked the beginning of a long tradition of ethnobotanical studies on the American continent. Therefore, this interest originated a tradition that will be fundamental for the future establishment, already in the 19th century, of a structuring and formalization of ethnobotany as a field of academic study.

If we go back in history, it seems wise to place the study of plants and their uses in the traditions of ancient Greek culture. These initial studies will blend the interest and formation of botany with ethnobotany, insofar as among the Greeks, botany seemed more like a study of medicinal plants. This beginning that blended the interests can be noted, for instance, in the writings left by Hippocrates, who wrote important works on the theme between 460 and 370 BCE (Baker, 1968).

Now, taking a leap in history, already in the 13th century, Marco Polo described an island producing pepper, nutmeg, and a host of other spices that can be found nowadays over the world. The report produced by Marco Polo spurred a search for the island that contained the spices. At that time, this endeavor was the necessary stimulus for the discovery of America by Europeans through the circumnavigation of the globe by Magalhães (Balick and Cox, 1996). Going to the time of Christopher Columbus, dated October 12, 1492, is his arrival in the Bahamas. From this and the next trips he would make, Columbus returned to Europe with plant species previously unknown to Europeans. Among these plants was tobacco (*Nicotiana* spp.), which was smoked by the natives on the island of Cuba. Other species that also crossed the ocean were corn (*Zea mays* L.), pepper (*Capsicum* spp.), for example (Cotton, 1996). What we can notice with this circulation of species from one continent to another is that, along with the plants, local observations were also sent about how the natives of the most diverse localities made use of these resources.

During centuries, including the period of discovery, postdiscovery, and colonization, the look at the uses of plants by natives was recorded by immigrants who arrived in the New World from Europe]. In this hall of people were the conquerors, adventurers, missionaries and other social actors. Specially from Portugal and Spain, these European documented in detail their observation of the knowledge and use of plants that mainly served as food and medicinal use for the original peoples.

With the circulation of plants, some of these new species introduced into Europe have been incorporated into the cultures of European countries and have become worldwide known and used until today. The discoveries of the flora of the New World produced commodities that are currently present in the world trade relationship. Adopting a universalist perspective, we can say that these early discoveries of ethnobotany align with economical botany and perfectly illustrate the importance of ethnobotany for the economic history of Europe. In another way, we can observe that occured the contrary movement of species as well as their associated cultural knowledge. That means that plant species used by local old societies from the so-called Old World were also introduced in this new geographical locations and gained a prominent position in the economic world relationships and world history. Althougth this historical fact of the first European contact with Amerindian societies has gained a world proportion in the ambit of the human constitution, this perception of the importance of species for human being has been experienced for a long time.

With an interest in the economic potential of plant resources considered exotic, during the first colonial period in the Americas, the study of ethnobotany was largely based on observations of local practices and evidences from anecdotes and chronicles. This ethnobotanical information was the basis of chronicles, diaries and other kinds of registers made by European, such as missionaries, conquerors, settler and other social actors of the 16th and later centuries.

Documentary production became diversified to the extent that this kind of information directed by unscientific eyes was complemented by documents that were precisely scientific in nature. This documentary typology has changed since the arrival of naturalists from Europe and the scientific expeditions organized in these new territories.

Immersing a little more into the question of plant function in a larger context, we can say that it was plants that played a major role in the European conquest project of the Americas (Crosby, 1986).

The circulation of biologic materials was a constant among the conquests, between the domains of the Portuguese, Spanish, and British Empires. Particularly from the second half of the 18th century onwards was established an extremely favorable period for the exchange of plants and botanical knowledge related to these plants. The political conditions favored a propitious moment to the articulation of a network of naturalists during the process of the conquests. This network of circulation of products and knowledge of natural history was composed by traveling naturalists, administrative agents, notaries, military officers, and traders (Medeiros, 2018). In this set of types of social actors, traveling naturalists went through the territories of Portuguese, Spanish, and English domains with the purpose of inventorying natural resources to send them to scientific institutions based in the European Empires (Pataca, 2006).

The relation between the sciences and the modern empires presents elements of how there was gradually a displacement of Renaissance conceptual paradigms to construct a natural history based on the new epistemes and practices of imperialism in the 18th century. With the turn of the 18th to the 19th century, the botanical gardens had already established themselves as a collection of plants from the most varied places of empires and as a portrait of the period of botanical discoveries of naturalists who were making a change in ideas about the world (Medeiros, 2018).

Already in the 19th century, naturalists such as Richard Spruce sent collections of specimens of plants and animals collected considered exotic to the museums and botanical gardens of Europe. Spruce was a prominent botanist who explored the northwestern Amazon and northeastern Andes, and he has been described as one of the greatest naturalists of all time (Schultes, 1983). Between 1851 and 1854, he circulated and explored the Rio Negro region in the Amazon. His scientific work focused on describing the ritual use of psychoactive species (Ford, 1978b).

In the same century American researchers in their own country, the United States of America, developed a great number of studies about the use of plants by indigenous. This new reality of North America in the production of scientific knowledge generated a change in the scientific sovereignty of Europe. With this, a new academic tradition was being formed and structured from the studies under a scientific perspective that later in the 19th century contributed to the conception of ethnobotany (Cotton, 1996).

3.2.2 Ethnobotany as a new academic discipline

Considering the second half of the 19th century, we can notice that studies about plants and their uses by native habitants of the Americas began to have a new structure and taxonomic descriptions of the flora of the American continent began to be more thoroughly studied. Another fundamental contribution to this scientific gain was the work of Edward Palmer and Stephen Powers, two US botanists who advocating for scientific rigor, they contributed largely to ethnobotany publishing many systematic treatment of collected data (Ford, 1978a). This process experienced by Powers would culminate in the introduction of the term aboriginal botany in 1873, referring to botanical researches concerning plants and their uses by natives. This term aboriginal botany would be used for the next 25 years in the academic field (Cotton, 1996).

Near the end of the 19th century, the term aboriginal botany and the interest in the subject began to widen. A significative mark in this process occurred at the 1893 World Fair. Preparations for the participation of the United States of America in this fair involved anthropologists and archeologists in the collection and in the preparation of traditionally useful plants that could be exposed (Ford, 1978a). This exhibition included the Hazzard collection, which comprised a range of preserved plant products used by the ancestors of the Pueblo indigenous group in Mancos Cañon in the state of Colorado (USA) and was sent to the University of Pennsylvania for analysis. At this university, the Hazzard collection was examined by botanist John Harshberger, who in his lecture in December 1895, gave for the first time in science the term "ethnobotany" when analyzing items in this collection such as clothing, agricultural tools, and other items from plants (Harshberger, 1896).

After this fact, the following decades were marked by an expansion and structuring of the studies in ethnobotany. A year after Harshberger's lecture, anthropologist Walter Fewkes introduced the newly coined term in anthropological literature by describing the etymology of the names of plants used by the indigenous of the Hopi ethnic (Fewkes, 1896). Following the timeline, the defense of the first doctoral thesis in ethnobiology occured in 1900 by David Barrows of the University of Chicago (USA) (Cotton, 1996). Thus, also in this beginning of the 20th century, in 1916, the redirects of Robbins and collaborators appear. Ethnobotany at this time changed its investigative field and began to include both the plants used by indigenous peoples and how the perceptions and understandings about them were reflected on varied cultures (Robbins et al., 1916). The latter approach was also addressed by the American ethnologist Melvin Gilmore, who worked on the necessity of having an in depth interpretation of the collected data considering its broader context, that means, cultural and linguistic aspects when making an ethnobotanical research (Gilmore, 1932).

With the intention of deepening ethnobotany on the part of scientists, data of different natures were compiled in an expressive way. For example, ethnology studies on plant uses in general were developed, and comparisons of species of particular uses by different human groups began to be made. The development of these studies brought contributions to the development of anthropologic theory, as well as to the delineation of ideas about cultural ecology and understanding of the human cognition (Conklin, 1954; Rappaport, 1968; Berlin et al., 1973).

Arriving in the 1980s, ethnobotany found academic recognition, especially in the United States of America. In 1981, the American Society of Ethnobiology was formed and published the first edition of its *Journal of Ethnobiology*. Following this event, there was the formation of the International Society of Ethnobiology - ISE (please, to find more information, a query on the topic can be made on the sites of *Journal of Ethnobiology* and ISE).

Since the late 1980s, articles have appeared in journals such as *Scientific American* (Cox and Balick, 1994) and *The Economist* (Anon, 1988), among others. Already by the 1990s, scientists also paid attention to the publication of the most popular books, which reached an unspecialized public, and whose publishing format and style we can find for sale with some ease even today. Thus, in the last two decades of the 20th century, the recognition of ethnobotany was growing both in the academic field as well as among the general public, who became aware of its existence through publications in the media where research data became more widely accessible to people.

Academic publications accompanied this process of expansion, and by the early 1990s, ethnobotanical study had grown promoting also an increase in the teaching process of ethnobotany. In the United States of America and the United Kingdom, postgraduate programs were available at universities in these countries. In the undergraduate courses of these same locations, students were brought to ethnobotanical science through the offer of disciplines such as economic botany and ethnobotany (Jain et al., 1986; Flaster, 1994). Other programs were also being structured between the 1990s and the 2000s in other regions of the American continent. The pedagogical discussion about the teaching of ethnobiology has been gaining ground in scientific meetings taking place in Brazilian territory. As a reflex of this process, in 2018 a book was released entirely dedicated to teaching practices in ethnobiology (Medeiros and Albuquerque, 2018).

Therefore, in this trajectory, we highlighted a new perception of the value of knowledge related to indigenous and traditional peoples and, thus, there was a new look at the maintenance of this knowledge that these peoples were the carriers.

Although there has occurred a shift in the production of scientific knowledge as we consider the relation between center and periphery, that is, between Europe and America, and particularly with regard to the development of ethnobotany we are referring to the United States, European ethnobotanists who have continued to bring important contributions to this field of research. European contributions have greatly enriched the discussions and deepened, for example, on ethnotaxonomy, traditional management in sustainable agriculture, ethnopharmacology, and palynology. These last two fields constitute the great contribution carried out by the ethnobotanists of Europe. The field of ethnopharmacology comprises the study of the scientific evaluation of medicines of local or traditional use. The field of palynology, as a strand of archeobotany, refers to the research of fossilized pollen, and it really gave an improvement in studies of paleoethnobotany. We can say that paleoethnobotany is an approach of ethnobotany that is dedicated to the study of past relation between people and plants from the perspective of the questions and methodologies that are proper to archeology. This is a kind of approach that comes close to historical ethnobotany (or ethnobiology) precisely because it considers the past history of the relations between people and elements of nature for an understanding of what we live today.

Immersing a little more in paleobotany studies, we have that some of the first scientists dedicated to this approach began to investigate the preserved pollen found in sedimentary deposits. This refers to the studies of Christian Ehrenberg, that during 1830 by immersing in this kind of perspective, suggested that these polen could indicate the history of the world's environment. In 1916, Norwegian geologist Lennart von Post developed a technique for quantitative pollen analysis in order to understand past changes in flora. Going forward in history, in 1941, another geologist named Johannes Iverson analyzed pollens relating them to archeological issues concerned to subsistence of people in prehistory at Barkaer's archeological site situated in northern Denmark. Through this investigation, Iverson managed to reach the history of civilization at the time when occurred the transformation of the hunter-gatherers to agriculture (Iverson, 1941). From the 1960s onwards, other palynologists in Europe, and also in the United States of America, used pollen analysis in works with archeological data to understand the anthropic action of the past on vegetation, on the environment (Dimbleby, 1963), which conferred to the paleoethnobotanical study greater expressiveness within the field of ethnobotany.

We have concentrated our historiographic analysis on the development of ethnobotany on the European and American continents, but of course the researches on knowledge of plants of traditional use have also had a history in Africa, Asia, and Australasia. In Africa, the traditional management of pastures and the use of wild and cultivated plants were the main themes in the investigations carried out (for example, Richards, 1985; Wickens et al., 1985; Abbiw, 1990). During its development, ethnobotanical research in both Asia and Australasia has encompassed studies of plants used in art, myths, and local practice in vegetation management. The first reports, or first disclosures of the data, maybe consulted in the publications of the Society of Ethnobotanists in India and in the Australian Institute of Aboriginal and Torres Strait Islander Studies in Canberra.

4. Brief synthesis

Closing this chapter, we can review the history of ethnobiology considering its approach through ethnobotany's development. Thus, the reflection on ethnobotany began in 1895 (Harshberger, 1896). When then it has been a term whose definitions have been accompanying the historical moments through which scientific development has passed in a broader context. It can be observed a conceptual evolution of ethnobotany through works which have been carried out and deepened. The interpretation of the term can be considered subject to change due to natural developments of the evolution of a scientific discipline. However, it is crucial to say that the essence of ethnobotany or ethnobiology as the study of interrelations between peoples and plants or biota has always been the basis of researchers' understanding. Through the time, ehtnobotany and ethnobiology have been and remain concerned with the register of peoples' cultural uses, perceptions and management of plants or elements of nature. So, not only the economic issues were the focus of the studies that have being carried out over time. In adition, to rescue part of the indigenous and traditional cultures the researchers have began to think about the evolutionary relation between human societies with plants or biota in general.

The successive layers of definitions that we have been able to follow in this chapter on the development that addresses the historical development of ethnobotany reflect the existence of varied opinions that find their origin in the multiple objects of the researchers involved in its study. Since its academic conception, ethnobotany has been developed by scholars with basic backgrounds in some scientific disciplines. In a concentrated analysis, what we can observe was that in the beginning, botanists focused their studies on Aboriginal societies and the economical potential of plants of cultural importance to them. Already, anthropologists focused their investigations on how different perceptions of natural environment could be very different between cultures and how these same perceptions could play a preponderant factor in making decisions for survival. Among these researchers were also found those who focused on oral culture and folk knowledge (Ford, 1978a). Throughout this chapter, this scientific history can be understood by analyzing some of the concepts of ethnobotany as an act of thought of the researchers, which has been summarized here.

As occured with its conceptual definition, the scope of ethnobotany has become broader encompassing botanical, anthropological and ecological aspects. Also, the students devoted to this field began to come from different academic backgrounds. The initially focus of the studies on the "economic" or "utilitarian" treatment is still present in ethnobotanical or ethnobiological studies. However, an emphasis on other issues it is perceived both in ethnobotany and ethnobiology, as commented before, pointing out that these sciences are essencially characterized for their interdisciplinarity and showing that there are many possibilities alongside difficulties in making research because there are many investigative interests and perpectives.

The strength of the development of ehtnobotany and ethnobiology was printed in periodicals specifically devoted to these scientific researches, such as the *Journal of Ethnobiology* and the *Journal of Ethnopharmacology*, but also in publications from different academic matrices, including, for example, archeology, paleobotany, phytochemistry, pharmacology, and conservation biology.

After dedicating ourselves in more detail to ethnobotany, we can conclude this chapter by focusing on the breadth of ethnobiology studies. Thus, the central idea that will prepare us to move forward in reading the next chapter is that the study of how a human group relates to its environment can be justified in itself. This statement is valid because this kind of investigation allows us to understand how people through their way of thinking and perceiving the natural environment exploit it, producing good living conditions and welfare for them (Godelier, 1981). Moreover, these studies open space for a discussion on the process of people's appropriation of nature to construct symbolic representations relevant to their social organization, memory and livelihoods (Posey, 1981, 1983, among others; Lévi-Strauss, 1970). Ethnobiology also brings elements to broaden the theoretical and methodological basis of understanding people's relation with natural environment and their mental structures to organize knowledge, memories and perceptions.

In addition, ethnobiologic research can bring us practical results based on the experience of the human group studied on a local, regional and global scales. These practical results would be related to the portion of knowledge held by indigenous and local societies, which can be appropriated and reframed in favor of a broader dimension of the society (whether regional or even global community). It is worth emphazising this insertion of local knowledge and practices for humanity, which is quite evident, for example, in medicinal plants and animals used as therapeutic agents. Therefore, different pharmacodynamic action derived from these species associated with local knowledge, promote the development of drugs and, from there, are incorporated into a global pharmacopoeia.

After immersing ourselves in the general understanding of the history of ethnobiology, ethnobotany, and the perception of the importance of time in the development of a scientific discipline, let's move on to the considerations of historical ethnobiology, a theme to be developed in the next chapter.

References

Abbiw, D.K., 1990. Useful Plants of Ghana. Intermediate Technology Publications Ltd/ RBG Kew, London.

Albuquerque, U.P., 2002. Etnobotânica para a conservação e uso sustentável da biodiversidade. In: Araújo, E.L., Moura, A.N., Sampaio, E.V.S.B., Gestinari, L.M.S., Caneiro, J.M.T., Orgs (Eds.), Biodiversidade, Conservação e Uso Sustentável da Flora do Brasil. Imprensa Universitária/UFRPE, Recife, pp. 244–246.

Albuquerque, U.P., 2005. Etnobiologia e biodiversidade. NUPEEA, Recife.

Albuquerque, U.P., 2013. Etnobiologia – bases ecológicas e evolutivas. Núcleo de Publicações em Ecologia e Etnobotânica Aplicada, Recife.

Albuquerque, U.P., Hurrell, J., 2010. Ethnobotany: one concept and many interpretations. In: Albuquerque, U.P., Hanazaki, N. (Eds.), Núcleo de Publicações em Ecologia e Etnobotânica Aplicada, Recife, pp. 87–102.

Albuquerque, U.P., Medeiros, P., 2013. Introdução à etnobiologia de bases ecológicas e evolutivas. In: Albuquerque, U.P. (Ed.), Etnobiologia: bases ecológicas e evolutivas. Núcleo de Publicações em Ecologia e Etnobotânica Aplicada, Recife, pp. 9–14.

Alcorn, J., 1995. Ethnobotanical knowledge systems: a resource for meeting rural development goals. In: Warren, D.M., Slikkerveer, L.J., Brokensha, D., Dechering, W.H.J.C. (Eds.), The Cultural Dimension of Development: Indigenous Knowledge Systems. Intermediate Technology Publications, London, pp. 1–12.

Amorozo, M.C.M., 1996. Um sistema de agricultura camponesa em santo Antonio do Leverger, Mato Grosso, Brasil. Ph.D. thesis. São Paulo University, São Paulo.

Anderson, A.B., Posey, D.A., 1989. Management of a tropical scrub savanna by the Gorotire Kayapó of Brazil. Adv. Econ. Bot. 7, 159–173.

Anderson, E.N., 2011. Ethnobiology: overview of a growing field. In: Anderson, E.N., Pearsall, D., Hunn, E., Turner, N. (Eds.), Ethnobiology. Wiley-Blackwell, New Jersey, pp. 1–14.

Anon, 1988. Just what the witch doctor ordered. Economist 307 (7544), 75–76.

Baker, H.G., 1968. Las plantas y la civilización. Herrero Hermanos Sucessores S.A., Ciudad de México.

Baleé, W., Gely, A., 1989. Managed forest succession in Amazonia: the Ka'apor case. In: Posey, D.A., Balée, W. (Eds.), Resource Management in Amazonia: Indigenous and Folk Strategies (Advances in Economic Botany 7). New York Botanical Garden, New York, pp. 129–158.

Balick, M.J., 1984. Ethnobotany of palms in the neotropics. In: Prance, G.T., Kallunki, J.A. (Eds.), Ethnobotany in the Neotropics (Advances in Economic Botany 1). New York Botanical Garden, New York, pp. 24–33.

Balick, M.J., Cox, P.A., 1996. Plants, People, and Culture – the Science of Ethnobotany. Scientific American Library, New York.

Barrera, A., 1979. La etnobotánica. In: Barrera, A. (Ed.), La Etnobotánica: Tres puntos de vista y una perspectiva. Instituto de Investigaciones sobre Recursos Bióticos, Xalapa, pp. 19–25.

Begossi, A., 1996. Use of ecological methods in ethnobotany: diversity indices. Econ. Bot. 50 (3), 280–289.

Berlin, B., Breedlove, P., Raven, P., 1973. The relation of folk systematic to biological classifications and nomenclature. Annu. Rev. Syst. Ecol. 4, 259–271.
Bodley, J.H., 1978. Preliminary Ethnobotany of the Peruvian Amazon. Report of Investigations of the Laboratory of Anthropology, Washington State University, No. 55, Pullman, Washington.
Cardona, G.R., 1985. La Foresta di Piume – Manuale di Etnoscienza. Laterza, Roma.
Castetter, E.F., 1944. The domain of ethnobiology. Am. Nat. 78, 158–170.
Clement, C.R., 1987. A pupunha, uma árvore domesticada. Ciência Hoje 5 (29), 42–49.
Clément, D., 1998. The historical foundations of Ethnobiology (1860-1899). J. Ethnobiol. 50 (3), 280–289.
Conklin, H., 1954. The Relation of the Hanuóo to the Plant World. Phd Thesis. Yale University, New Haven.
Cotton, C., 1996. Ethnobotany: Principles and Applications. John Wiley & Sons, Chichester.
Cox, P.A., Balick, M.J., 1994. The ethnobotanical approach to drug discovery. Sci. Am. 270 (6), 82–87.
Crosby, A.W., 1986. Ecological Imparialism: The Biological Expansion of Europe 900–1900. Cambridge University Press, Cambridge.
De Candolle, A.P., 1819. Théorie élémentaire de la botanique: ou, Exposition des príncipes de la classification naturelle et de l'art de décrire et d'étudier les végétaux. Chez Deterville, Paris.
Dimbleby, G.W., 1963. Pollen analyses from two Cornish barrows. J. R. Inst. Corn. 4, 364–375.
Elisabetsky, E., 1986. New directions in ethnopharmacology. J. Ethnobiol. 6 (1), 121–128.
Fewkes, J.W., 1896. A contribution to ethnobotany. Am. Anthropologist 9, 14–21.
Flaster, P., 1994. SEB classes. Plants People Soc. Econ. Bot. Newsl. 2. Spring.
Ford, R.I., 1978a. Ethnobotany: historical diversity and synthesis. In: Ford, R.I. (Ed.), The Nature and Status of Ethnobotany (Antropological Papers, Museum of Anthropology, University of Michigan No. 67). Ann Arbor, Michigan, pp. 33–50.
Ford, R.I., 1978b. Prehistoric Food Production in North America. Museum of Anthropology, University of Michigan, Ann Arbor.
Ford, R.I., 1986. An Ethnobiology Source Book: Tehe Use of Plants and Animals by American Indians. Garland Publishing, New York.
Gilmore, M.R., 1919. Uses of Plants by the Indians of the Missouri River Region. University of Nebraska Press, London.
Gilmore, M.R., 1932. Importance of ethnobotanical investigation. Am. Anthropol. 34, 320–327.
Godelier, M., 1981. Godelier: Antropologia. Ática, São Paulo.
Hardison, P., Bannister, K., 2011. Ethics in ethnobiology: History, international law and policy, and contemporary issues. In: Anderson, E.N., Pearsall, D., Hunn, E., Turner, N. (Eds.), Ethnobiology. John Wiley & Sons, Inc., pp. 27–49
Harshberger, J.W., 1896. The purposes of ethno-botany. Am. Antiqu. 17 (2), 73–81.
Hunn, E., 2007. Ethnobiology in four phases. J. Ethnobiol. 27 (1), 1–10.
Irvine, D., 1989. Succession management and resource distribution in an Amazonian rainforest. In: Posey, D.A., Balée, W. (Eds.), Resource Management in Amazonia: Indigenous and Folk Strategies (Advances in Economic Botany 7). New York Botanical Garden, New York, pp. 223–237.
ISE – International Society of Ethnobiology, 2014. http://ethnobiology.net/. Accessed 02.06.2019.
Iverson, J., 1941. Land occupation in Denmark's stoneage. Den. Geol. Unders. 2 (66), 1–67.
Jain, S.K., 1981. Glimpses of Indian Ethnobotany. Oxford & IBH, New Delhi.
Jain, S.K., 1987. A Manual of Ethnobotany. Scientific Publishers, Jodphur.

Jain, S.K., 1989. Methods and Approaches in Ethnobotany. Proceedings of the 2nd Training Course in Ethnobotany, Lucknow. Surya Publications, Dehradun.
Jain, S.K., Minnis, P., Shah, N.C., 1986. World Directory of Ethnobotanists. Surya Publications, Dehradun, India.
Jones, W., 1941. The nature and scope of ethnobotany. Chron. Bot. 6 (10), 219–221.
Lévi-Strauss, C., 1970. O Pensamento Selvagem. Nacional, São Paulo.
Marques, J.G.W., 2001. Pescando Pescadores: ciência e etnociência em uma perspectiva ecológica. NUPAUB, São Paulo.
Martin, G., 1995. Ethnobotany: A Methods Manual. Chapman & Hall, London.
Medeiros, M.F.T., 2009. Etnobotânica histórica: princípios e procedimentos. Núcleo de Publicações em Ecologia e Etnobotânica Aplicada, Recife.
Medeiros, M.F.T., 2018. Medicinal plants and state policy in South America: the case of colonial Brazil. In: Albuquerque, U.P., Patil, U., Mathe, A., Orgs (Eds.), Medicinal and Aromatic Plants of South America - Brazil. Springer, Dordrecht, pp. 71–80.
Medeiros, M.F.T., Albuquerque, U.P., 2018. Práticas para o ensino de etnobiologia. Núcleo de Publicações em Ecologia e Etnobotânica Aplicada, Recife.
Pataca, E.M., 2006. Terra, água e ar nas viagens científicas portuguesas (1755-1808). Ph.D. thesis. Caminas State University, Campinas.
Phillips, O., Gentry, A.H., 1993. The useful plants of Tambopata, Peru: I. Statistical hypotheses tests with a new quantitative technique. Econ. Bot. 47 (1), 15–32.
Posey, D.A., 1981. Ethnoentomology of the Kayapó Indians of Central Brazil. J. Ethnobiol. 1 (1), 165–174.
Posey, D.A., 1983. Indigenous knowledge and development: an ideological bridge to the future. J. Ethnobiol. 35 (7), 877–894.
Posey, D.A., 1986. Introdução – etnobiologia: teoria e prática. In: Ribeiro, B.G., Coord (Eds.), Suma Etnológica Brasileira – 1. Etnobiologia. Vozes, Finep, Petrópolis, pp. 15–251.
Posey, D.A., 1987. Etnobiologia y ciencia "folk": su importancia para la Amazonia. Hombre y Ambiente 1, 7–26.
Powers, S., 1873. The Californian Indians. Overl. Mon. Out West Mag. 10 (4), 322–333.
Prance, G.T., 1991. What is ethnobotany today? J. Ethnopharmacol. 32, 209–216.
Rappaport, R.A., 1968. Pigs for the Ancestors: Ritual in the Ecological of a New Guinea People. Yale University Press, London.
Richards, P., 1985. Indigenous Agricultural Revolution. Unwin Hyman, London.
Robbins, W.W., Harrington, J.P., Freire-Marreco, B., 1916. Ethnobotany of the Tewa Indians. Bur. Am. Ethnology Bull. 55.
Robertson, R., 1992. Globalization: Social Theory and Global Culture. Sage, London.
Schultes, R.E., 1962. The role of the ethnobotanist in the search for new medicinal plants. Lloydia 25 (4), 257–266.
Schultes, R.E., 1972. De plantis toxicariis e Mundo Novo tropicale commentationes X. New data on the malpighiaceous narcotics of South America. Bot. Mus. Leafl. 23, 137–147.
Schultes, R.E., 1983. Richard Spruce: na early ethnobotanist and Explorer of the northwest Amazon and northern Andes. J. Ethnobiol. 3, 139–147.
Schultes, R.E., Reis, S. von, 1995. Ethnobotany: Evolution of a Discipline. Timber Press, Portland; London.
Toledo, V.M., Alarcón-Cháires, P., 2012. La Etnoecologia hoy: panorama, avances, desafios. Etnoecologia 9 (1), 1–16.
Toledo, V.M., Barrera-Bassols, N., 2009. A etnoecologia: uma ciência pós-normal que estuda as sabedorias tradicionais. Desenvolv. Meio Ambiente 20, 31–45.
von Post, L., 1916. Einige Südschwedischen Quellmoore. Bulletin of the Geological Institution of the University of Upsala 15, pp. 219–278.
Wickens, G., 2000. Defining economic botany and ethnobotany. Econ. Bot. 54 (3), 251.

Wickens, G.E., Goodin, J.R., Field, D.V. (Eds.), 1985. Plants for Arid Lands. Unwin Hyman, London.

Wolverton, S., 2013. Ethnobiology 5: interdisciplinarity in an era of rapid environmental change. Ethnobiol. Lett. 4, 21–25.

Wolverton, S., Chambers, K.J., Veteto, J.R., 2014. Climate change and ethnobiology. J. Ethnobiol. 34 (3), 273–275.

Xolocotzi, E.H., 1983. El concepto de etnobotánica. In: Barrera, A. (Ed.), La etnobotánica: três puntos de vista y una perspectiva. Instituto Nacional de Investigaciones sobre Recursos Bióticos, Xalapa, pp. 13–18.

Xolocotzi, E.H., 1987. Xolocotzia. Tomo II, Rev. de Geografia Agricola. Universidad de Chapingo, Mexico.

CHAPTER TWO

Defining historical ethnobiology

Contents

1. Theoretical matrix and concept of historical ethnobiology

Considering a theoretical matrix for historical ethnobiology compels us to think about the function of "things" (objects/symbols) concerning people. Based on the relations established between people and "things", the fabric of the theoretical framework of this scientific field is established. This is so because the objects/symbols awaken in the individual's memories, events, people, and values that are to be conserved and transmitted according to a criteria selection and determination of their importance for the survival of people.

Internal memory is constituted internally through images to represent the particular collection of material objects and immaterial information intended to activate and provoke memory reactions.

Memory is continually constituted through the preservation of elements and the establishment of relationships. It fixes or records data from human perception, experience, and knowledge. However, it does not stagnate in the conservation of objects, images, and information. When memory is activated, it brings to life everything that has been preserved by it.

By associating data from material and immaterial images, memory promotes different articulations between its elements. This process happens along with imagination. Hence, memory and imagination dialogue over time insofar as they are not limited to the present moment. This is why external memory records are important since they enable the expansion of information repositories that can be accessed at any time and place. Memory precisely allows this access and the retrieval of something that is from the

Historical Ethnobiology
ISBN: 978-0-12-816245-3
https://doi.org/10.1016/B978-0-12-816245-3.00002-0

past, which belongs to the past. Through this action, past data can be compared or related to the information in the present.

The importance of memory lies, therefore, in the ability to preserve essential information for the elaboration of individual and collective experience at various levels, such as the scientific, philosophical, and everyday knowledge.

In their path toward memory, ideas and words can be grouped, and facts can assume their direction in time and social continuity. The reverse of this memory frame would be the absence of social links, individual and collective identities, a notion of society, and also the possibility for the nonexistence of knowledge. In this sense, memory is the foundation for cultural transmission.

The primordial forms of circulation of ideas between people have long been through the constitution of images, the correct use of orality, and the written word (Belting, 2004). It is interesting to note that in nonhistorical societies, memory is formed through timeless recall, and it is related to the myths of their ancestors. In societies where writing is in existence, history can be chosen by adopting a chronologic and documental reading. It is observed that in case of presence or absence of writing, memory is an imperative condition among human societies (Yates, 1993; Candau, 2005).

For the preservation of ideas, societies started developing forms of oral memory elaboration, usually kept by older members, who became living records of the knowledge chosen as being essential to the survival of the group. In this way, customary laws, the foundations for coexistence and traditions are elaborated in forms of mythical narratives.

Oral narrative is formed as a regular discourse to make its memorization possible. In societies where writing is nonexistent, the memorization of words and/or concepts is not a mechanical act learned through rote techniques. The oral narrative form is a mnemonic act, which exerts the social communication role and conservation of information in the absence of the object narrated and the impossibility of writing (Rossi, 2004).

The endeavor for the preservation of symbols, languages, and objects that are the base for people's collective memory is a concern in all human societies. What has changed over the course of history has been the techniques available for registration and preservation of information. Therefore, technologic changes have altered the structures on which safeguarding and social sharing was intended to lie.

By moving from oral to written record, humanity managed to increase information storage from individual memory to other memories. The preservation of oral memory in the form of written record representation also limited information loss. This process occurred insofar as the demise of the memory bearer ceased to be the cause of memory element loss kept by such a person.

The principles of memory are the relationship between order, place, and images. In relation, these principles allow for the retrieval of things, concepts, and words stored in people's memories.

Orality transfer to writing has created a new form of collective memory. In this process of production of written vestiges, the memorization has gained a precious space. Through this mind organization of the information it became to be formed a set of words, expressions and concepts. These sets started to be systematized in the mind, following an order of organization of these vestiges that allow a contextualization of the information when passed on to a specific support.

If we now no longer turn our gaze to the documental production of the past, and we turn to the technology available today, the scenario changes. With the advent of information thechnology, the new devices started to discipline individual and collective memory, becaming the mediators between the user and the infortmation represented by memory. Therefore, the new technologies have brought along an "issue" that the area of historical ethnobiology has to tackle. To some extent, this issue is set on account of the difficulty defining the field and object posed by the technologic revolution going on in the last decades. Research sources have become integrated with other documental types until then inexistent.

Here lies the ethnobiologist's skill: to know how to organize information means to know how to choose and associate based on enunciation and discourse logic. The organizing principles are not ready-made, and due to technologic changes, they are continually changing. Moreover, rather than being available *ad lib*, these principles stem from historical formations geared to the needs of actors and social groups.

At this juncture, it is time we thought about the conceptualization of the term "history" (historical) that composes this ethnobiology bias. History comprises the gathering and study of knowledge documented or transmitted by tradition, concerning the development of humanity, art, or science, a particular period, people, region, or individual (universal history, history of medicine). It is also the succession of events or facts that characterize an action, and also the discipline, science, and methods of this study (Ferreira, 2010).

In this historical approach perspective, the study of documental evidences would be one of the paths available for the investigation of multiple aspects of the relation between the elements of nature, cultures, and human societies.

Nowadays, access to written documents, books, and scientific articles that deal with historical accounts is readily available. These materials show the interest endowed to living beings and their bond with people.

In the developmental history of ethnobiology, Melvin Gilmore, in 1932, and Volney Jones in 1941 redirected research, particularly in ethnobotany. These US researchers ceased restricting studies to listings of popular plant names with their respective botanical identifications and local uses. They played an important role in historical ethnobiology, as they came to recognize the study of past cultures. Starting with them a historical approach for the study of a direct interrelationship between people and plants has been made possible (ethnobotany).

With the development of the studies from these researchers in the first half of the 20th century, research began to contemplate both the material and immaterial evidence preserved in the most varying cultures. Gilmore and Jones also endeavored to consider this evidence together with information on the ecologic characteristics of vegetables. Furthermore, they incorporated the quantitative analysis of knowledge and use of plants in their scientific practice. The focus of Gilmore and Jones's study was to seek answers to questions concerning changes in eating habits, forms of subsistence, and how plant species domestication originated (Hastorf and Popper, 1988).

When we think of the contributions from material evidence, we can classify them as being unique and/or shared between paleoethnobiology and historical ethnobiology. See Table 2.1 for the distribution of evidence between these approaches.

The constitution of a document *corpus* for historical ethnobiology, although still open and unfinished as to its amenability as a documental type, is enclosed in the expression of the thought of the authors who left

Table 2.1 Discrimination of own uses evidence and shared evidence between paleoethnobiology and historical ethnobiology.

Research	Evidence	Research	Evidence
Paleoethnobiology	Plant and/or animal fossil Evidence of human remains (indirect evidence)	Paleoethnobiology/historical ethnobiology	Objects (clothing, artifacts, etc.)
Historical ethnobiology	Iconographic evidence Photographic evidence Written evidence Objects		Iconography (cave paintings and paintings on decorative items, etc.)

behind textual and imagery evidence of their own time. Such evidence deals with the social memory of human societies; they matter to science and must be valued as cultural heritage.

The analysis of the informational attributes of such evidence is in the realm of historical ethnobiology. That is what it stands for. Therefore, it encompasses the study of the interrelations established between people and the elements of nature, over time and in varying environments, these interrelations being determined by ecologic and cultural conditions. It is the current understanding of how biodiversity was discovered in the past.

The research process in historical ethnobiology assumes this retrospective look, and it is concretized, precisely, through the study of historical evidence. It contains information evidence on knowledge and people's practices in relation to the elements of nature in regard to the most varied living environments from times past. In this investigative perspective, we make an attempt to find the "how" and "why", to strive to describe, characterize, understand, articulate, and recreate the role nature played in past human societies, and how they function at the present time, as well.

With the development of historical ethnobiology, we intend to reach some objectives. It is important to objectively describe the properties of any element of nature and how they are inserted in our lives. Still another objective in this area is to find out about species that have played important roles in social, cultural, and economic issues in the past (Medeiros, 2007). If we keep on looking solely to information preserved in the past, a last objective would be to fill in a few gaps in the understanding of how certain species were present in various cultures.

Furthermore, historical ethnobiology research can be performed by taking a retrospective view on information concerning the species studied. By following this research path, we will be able to compare past practices and knowledge with current understanding. Such comparisons can take place through the analysis of different popular names of plants, animals, landscapes, or other elements of nature. Also, a comparison analysis of orthographic variations between different historical epochs could be performed (past and present time). To try to understand the cultural significance of a given species, an analysis of past and actual uses, preparation modes, and the parts utilized, and geographic occurrence should take place. Yet another research in this perspective is to examine all possible new applications for both native and exotic species.

Still another important role historical ethnobotany plays is in the analysis of an author's viewpoint, intent, and posture upon producing an original document (Albuquerque, 1997; do Nascimento and de Carvalho, 2006). The author of a document is embedded in the culture of his/her time.

Therefore, this person is responsible for the prepared information source and is supposed to have a massive knowledge about all kinds of elements of nature in respect to the limitations and confines of a given era. For that reason, the researcher dedicated to the examination of historical sources must not solely focus on the document itself. That said, a good research practice in historical ethnobiology would be to try and grasp a more encompassing context from which the document was produced.

Pursuant to this view, a few questions are central in developing research that is committed to finding details concerning the sources under examination. A few of these are as follows: Who was the author? Who sponsored the work? What was the sociocultural context in which the document was produced? And, what theories and practices were usually accepted when the document was produced?

It is paramount that the researcher should be able to respond to these questions so that interpretations of past evidence collected from these documents are supported by ecologic, social, economic, and cultural contexts of the epoch in which they were fashioned. This type of certification and treatment of the sources will help guarantee that the analyses resemble closely the reality of the past time now under analysis. Moreover, the risk of an anachronistic understanding of the documental source can be greatly reduced.

The exercise in examining the context of any documental production with a broad perspective is what makes the creation of a series of debates on the sources' time of origin possible. This kind of posture also allows for an in-depth examination and consideration of the ideas of the past time in which the documents were produced.

2. Related knowledge areas to make possible historical ethnobiology research

To trace the encompassing field of historical ethnobiology is to recognize, as a basic condition, a reflection that establishes orderly relationships between nature and human societies. It is important to assert that it is a relatively autonomous branch of ethnobiology. This also implies that there is an interdisciplinary condition. This structuring characteristic takes place through the relation between scientific fields, whose essential set is specially formed by the areas of biology, history, archiveology, and museology.

The ethnobiologist striving to work in this historical development field is supposed to elaborate mainly narratives, although not exclusively, from a written memory. In this sense, the whole first part of this chapter, and what will be further elaborated in the next chapter (Chapter 3) will elicit meaning provided the ethnobiologist becomes inserted in a perspective that prompts him the questions on what memory is like, what it stands for, and how it works.

Writing and reading as access to memory represent more than just a form of learning and knowledge production. Rather, they mean the possibility of a cumulative communication that is revealed as essential to the scientific order and, consequently, to historical ethnobiology.

The ideas presented so far stem from the proposition that the object—culture—memory focus is interdisciplinary in nature. In this view, we put independence and, at the same time, interdisciplinarity as structuring in the development of research on historical ethnobiology. Therefore, theories and methodologies or sets of technical procedures for ascertaining a given discipline are not sufficient for conducting research with multiple approach characteristics.

We must not, however, fail to mention that the areas of knowledge that go beyond the fundamental fields already discussed can also integrate the theoretical-methodological framework of research. This determination of the inclusion of other areas is linked to the scope of the theme, which will be worked on, and will be dependent on a definition, by the ethnobiologist.

3. Themes of interest of historical ethnobiology

Throughout the history of human civilizations, many historical printed and manuscript sources along with other types of objects were produced by people. We will comment in more detail about the sources of these documents in the following chapter. For the time being, we need to mention that these documents constitute valuable research tools concerning traditional knowledge and biodiversity.

By thinking about the information preserved in these documents in a general manner, we can point out the themes of interest in historical ethnobiology. We can also affirm that preserved information concerning traditional cultural-biologic knowledge can be found in these documents. Moreover, these sources can reveal the cognitive complexity of past eras accessed nowadays in the form of registered myths, stories, and symbolic representations in the arts and in ceremonies (Medeiros, 2010).

However, it is safe to say that the themes addressed in the documents produced by humanity are extremely broad in terms of the actors who produced them and their interests. If we were to draw a profile of textual production by people over time, making a cut from themes of interest to historical ethnobiology, we could say that the initial focus of documental production was geared to the economic application and potential of plants and animals used by native peoples. Anthropologic aspects were also considered important to register. Taking this set as base for documents, which are directed to the registry of the natural world and its relation with people, we can point out a direct way to the subjects with oftener presence in documental production. These themes will revolve around medicinal and edible plants and animals, timber and dyeing plants, hunting techniques, characteristics of the local culture and its people, and, consequently, cultural and biologic diversity. Also, themes like cultivation techniques, domestication of species, and the distribution of species, agricultural practices, and description of the landscape are often present in the descriptions.

References

Albuquerque, U.P., 1997. Etnobotânica: uma aproximação teórica e epistemológica. Revista Brasileira de Farmácia 78 (3), 60–64.

Belting, H., 2004. Pour une anthropologie des images. Gallimard, Paris.

Candau, J., 2005. Anthropologie de la mémoire. Armans Colin, Paris.

Ferreira, A.B.H., 2010. Dicionário Aurélio da Língua Portuguesa. Editora Positivo, Curitiba.

Hastorf, C.A., Popper, V.S., 1988. Current Paleoethnobotany – Analytical Methods and Cultural Interpretations of Archeological Plant Remains. The University of Chicago Press, Chicago/London.

Medeiros, M.F.T., 2007. Fontes documentais do século XIX: fundamentos para a pesquisa etnobotânica hodierna. In: Barbosa, L.M., Santos Júnior, N.A. (Eds.), A botânica no Brasil: pesquisa, ensino e políticas públicas ambientais. Sociedade Botânica do Brasil, São Paulo, pp. 565–568.

Medeiros, M.F.T., 2010. Historical ethnobotany: an approach through historical documents and their implications nowadays. In: de Albuquerque, U.P., Hanazaki, N. (Eds.), Recent Developments and Case Studies in Ethnobotany. Núcleo de Publicações em Ecologia e Etnobotânica Aplicada, Recife, pp. 127–142.

do Nascimento, D.R., de Carvalho, D.M., 2006. Apresentação. In: do Nascimento, D.R., de Carvalho, D.M. (Eds.), II Seminário História das Doenças. FIOCRUZ, Rio de Janeiro.

Rossi, P., 2004. A Chave Universal: Artes da memorização e lógica combinatória desde Lúlio até Leibniz. Edusc, Bauru.

Yates, F.A., 1993. L'Arte Dela Memoria. Einaudi, Torino.

CHAPTER THREE

Documents that reveal the interactions between people and nature

Contents

1. The logical meaning of the terms involved in the documentary analysis for ethnobiology

The theme addressed in this chapter points to the following questions: Is a historical ethnobiology possible? If we admit this possibility, can we articulate remembrance and oblivion in this perspective? Finally, would not memory be an inflection point, the arc of dialogue between the constituent elements of historical ethnobiology?

To this inquiry, another uneasiness was overtaken: thinking the memory. Restlessness that was not about searching concepts of memory, but rather to establish an angle that would allow us to see and question how the relationship between memory and ethnobiology occurs.

Historical Ethnobiology
ISBN: 978-0-12-816245-3
https://doi.org/10.1016/B978-0-12-816245-3.00003-2

Let us say some words about memory. Memory has been considered a source of knowledge since classical antiquity. Along the time, the areas of knowledge of philosophy, psychology, and psychoanalysis have contributed to the reflection about memory aiming to better understand human mind and its intimate manifestations. So, it has been a subject of reflection and questioning. Thus, following the standard of the modernity conduct present in contemporary society, memory has always been resized.

Among the aspects that involve this new perception of memory, some deserve to be commented on. The first postulates that memory has a singular dynamic when we realize that it is not only remembrance but also oblivion that together play the balance between oppositions and complementarities. That is to say, that the existence of people in relation to environments, specific or general situations, and certain places are alive in the memory because they take the place of some other records that have been forgotten. This is the balance of memory.

The second aspect brings us to the idea of the existence of both individual and collective memory. What is remarkable in the memory is the dualism between *ratio et adfectus* (*reason and emotion*). A set of experiences manifests itself in the individual in a multifaceted way. Different reflections and senses of an individual memory produce a collective memory in a society, in a group, in a collectivity. On the other hand, the collective memory influences the individual memory, since it permeates the individual memory of those who live in such a group. Therefore, there is a constant interaction between individual and collective memory, thus constituting a general memory (Murguia, 2010).

Having in mind the framework of the symbolic and imaginary, we point to memory as a phenomenon in the trace. That is to say that through the records, the memory gains its materiality. Thus, it is opportune to think about the definition of some constituent terms of memory, which we will present later.

Traces or vestiges are social memory documents when recalled in the collective dimension. Thus, what we call memory can be comprehended as the great arsenal of all that we live (individually and collectively) and that it could be interesting to remember (Namer, 1987).

Given these considerations, would not it be appropriate to think of social memory as being strongly represented by cultural heritage? Let us then go on to discuss some considerations on "culture", "heritage", and "object".

By well placing the relation of these terms with the social memory, Dodebei (2005) gives us a conception of them. "Culture" gives the structure to social memory, and its existance points that the records are alive. "Heritage" encompasses the deposit of all records which express the meanings and feelings to the memory. "Object", intrinsic to the heritage, can be expressed by textures, sounds, places, persons, and representations both in the material or virtual universe, and it serves as an articulator of social thougths, actions and relations. The biocultural heritage, as defined in the Code of Ethics of the International Society of Ethnobiology (ISE, 2006), "is the cultural heritage (both the tangible and intangible including customary law, folklore, spiritual values, knowledge, innovations and practices) and biological heritage (diversity of genes, varieties, species, and ecosystems provisioning, regulating, and cultural services) of Indigenous peoples, traditional societies, and local communities, which often are inextricably linked through the interaction between peoples and nature over time and shaped by their socio-ecological and economic context. This heritage includes the landscape as the spatial dimension in which the evolution of Indigenous biocultural heritage takes place. This heritage is passed on from generation to generation, developed, owned and administered collectively by stakeholder communities according to customary law".

Considering the fact that throughout history we can observe a social desire of having all preservated, these conceptualizations about culture and cultural heritage have moved to a broader understanding of their meanings. The scope of the concept of cultural heritage followed the research developed by anthropologists for the idea of culture (Van-Mensch, 1992).

In this sence, these researches created a gradient of humanity for establishing patterns of humans societies considering the culture as the main category of this evolutionary proposal (Abreu, 2008). In this evolutionary conception, culture was valued through the manifestation of human beings' thoughts and actions. Observation and collection of information and objects through fieldwork became instruments through which anthropologists could propose patterns of similarities and distinctions between different cultures. When developing fieldworks, a enourmous amount of objects started to form the necessary evidence to verify these cultural traces and the collections of museums. But, why collect these objects?

Collecting them is important for several reasons. This way, the objects of the indigenous and local community are manufactured with materials from plants, animals or other elements of nature. These culturally produced materials include, for example, musical instruments, fabrics, food, medicines, household utensils, hunting and fishing guns and traps, and so on. Thus, these objects are transformed into a document by an act of investigation when considering them a representation of a unique culture. Through this transformation of objetcs into documents, they lead us to a profound conception that tells us about an epoch, a person, the economy, and the relation between human communities and their living environment. In addition, they represent an origin memory which shows us their cultural values and craft techniques of the local people.

Then, with the development of research in the anthropologic area, human societies passed to attend a gradient based on the concept of culture within a classification of patterns of similarities and differences between human societies, when considering cultural or social anthropology.

Going to a historical context and considering the American continent, since the 16th century, the interest about plants, animals and other elements of nature considered exotic and their uses and acclimatization was part of a political, economic and scientific project of European colonization. In these colonization travels, the Europeans could collect materials that would supply their offices of curiosity, botanical gardens and also museums.

Although this interest has long been present in the history of human civilization the dynamics of the exploration of nature on the part of European empires expanded greatly during the subsequent centuries before the discovery of the New World through a project of colonization, specially in the last decades of the 18th century. Thus, in this process of discovering so many new things, the area of making collections was being established as a manner of collecting and maintaining together objects of the cultural heritage. So, through this practice there was a gradual recognition of the importance of the existence of botanical gardens and natural history museums, specially in Europe.

At that time, the European universities were rethought from the execution of the Illuminist reform of the studies, which translated into a decided bet in the natural faculties, necessarily supported by establishments able to combine theory with practice, introducing in the learning process observation and demonstration. For this became the indispensability of a botanical garden that could house the practical classes in the area of botany. A concrete example of this evolution of the curiosity offices, botanical gardens, and the teaching of natural sciences can be reported through the work of Domenico Vandelli, an Italian physician and naturalist who, to the invitation of the Marquis of Pombal, worked in Portugal between 1764 and 1815. Vandelli was distinguished in the preparation and execution of the Illuminist reform to the studies of the University of Coimbra (Portugal) and collaborated actively in the creation of the botanical garden in this educational institution.

To complement the Office of Natural History, the Statutes of the University of Coimbra, 1772, advocated the indispensability of a botanical garden, dictating, first and foremost, the objectives to which this space should obey, as can be seen in the following section:

"Although in the Office of Natural History were included the productions of the vegetable kingdom, plants cannot be seen in it, but in their cadavers, dry, macerated, and embalmed; it will be necessary to complement the same history in the establishment of a botanical garden in which live plants are shown. Therefore, in its place, which is more appropriate, and competent in the vicinity of the university, it will soon establish the said garden: to cultivate all the genus of plants, especially those of which one knows or if one expects any service in medicine and in the other arts, with the necessary care and providence to assemble the plants of my overseas domains, which have immense riches, in what belongs to the vegetable kingdom" (Statutes of the University of Coimbra, 1772, pp. 266–267).

A multifaceted 18th-century character, Vandelli is in the center of the process of developing an imaginary of the European natural history museums. The collections of this spaces began to be established through the exercise of donations and benefits. In this sense, with the accumulation of the donations of materials collected and their perservation in the offices of curiosity it was started the formation of the collections, that means to say that the social memory began to be represented in each object of the collections. At this point, museums are understood as special spaces, in the words of Wagensberg (2005), a space where there is a concentrated reality.

In a science museum, an object begins to represent a range of possibilities, which includes the representation of an individual, a genre, a class, a family of objects, as well as distant places, phenomena, or objects impossible to be displaced or inaccessible to human perception, because they are infinitely small, infinitely large, or dispersed in time and/or space. A museum object can be of artificial or natural nature, be it dead or alive, human, animal, floristic, mineral, organic or inorganic, and also unique or representative. Museum objects, separated from their original context, that is, its primary context, are put in relation into a new reality represented by the space of the museum as a manner of transforming these objects in a document that reveals the memory of the reality from which they came from (Van-Mensch, 1992). In this way, as the objects represent a whole historical past, they are considered as national goods by documenting an original memory of the reality from which they came (Breffe, 2005; Van-Mensch, 1992).

Gradually, museum collections are being formed. The collections of museums are the hierarchic organizational units. They preserve these objects, which are of different types, natures and origins. Also, they have a characteristic to propose different queries through narrative strategies, and to confer materiality and visibility to intangible realities to human perception. Thus, if reality cannot be collected and preserved in its entirety by museums, they can, through their collected, reorganized, and recontextualized vestiges (the objects) assign visibility to these inaccessible realities (this includes, for example, events, phenomena, scientific concepts, etc.).

In turning to the universe of objects and material culture, museums express mechanisms of memory institutionalization. So, museums represent the possibility of constitution of a totality by the process of ordering and classificating vestiges that were collected putting them in relation to each other by proposing a coherent system. Vestiges displayed in a museum remind us of a reality that is necessarily larger and more complex. This way, objects serve as means for the proposition of a narrative, an act of thought, and for the reconstitution of aspects that involved the *modus vivendi* of human societies, their livelihood environment, their simbolic representations, and so on. Meneses (1992) emphasizes that the musealization axis is the research process of transforming the object into a document that brings in its constitution many references of other spaces, epochs, and meanings in a contemporary reality of museums, collections, exhibitions and their users.

In this context, we must reflect on what historical documentation stands for. The documents are the set consisting of the information and its support, which can be used for consultation or proof. The support consists of the material that serves as a basis for the different forms of recording information, such as film, paper, magnetic tape, disk, etc. They can be divided into two classes: iconographic and textual documents. The first are archival documents presented in the form of a printed, handwritten, or photographic image. They are documents such as the original and/or reproduction of artwork, photography, technical drawing, slide, transparency, and poster. Textual documents are documents with written records, in different supports. They include typed or printed manuscripts.

From the adoption of positivist thinking and the change in thinking about the questions specifically related to natural sciences, teaching, botanical gardens, and zoological collections, it was during the middle of the 19th century that the botany economic collections, as were called the biocultural collections at the time, began to be classified as such. This conception about the collections was developed during the expansion of the European empires. The conquest and domination plan outlined by the Europeans had two main motivations: to gain the international prestige for having these collections and to maintain and expand the business relations by providing the industrialized economies with raw materials. What happened in Europe was the creation of botanical gardens and ethnographic museums, which safeguarded collections of living and dry plants and objects, and collections of "primitive societies" artifacts. These botanical gardens, ethnographic and natural museums were created with the aim of stimulating and supporting research on the biological sociocultural and biocultural diversity of the world.

Thus, collections of plants, animals and other elements of nature and cultural objects from indigenous and other traditional societies and local populations were related to the scientific and economic aims that guided the exploration travels of the European empires. The process of formation of these collections were established through these travels and the preservation and access to them in the museums and botanical gardens were made through donations, orders, and sales. The collected items were sent to these institutions by social actors that played an important role for these projects of domination and expantion. This effort was carried out by settlers, travelers, naturalists, religious, militaries, rulers, and later in history, by scientists as well as by other social actors. In these memory institutions, collections of plants, animals, minerals, and artifacts have been documented and preserved for centuries. This way, these items from the past reach the present moment, when they become sources of consultation and a possibility of study for the researchers of today.

As we can see, the expansion of the idea of cultural heritage accelerated rapidly over time. The common idea is that we must preserve all the repertoire that we inherited, or we can lose it forever. Today the notion of cultural heritage is of great scope, what it means to say that it encompasses all the things, if it were possible to us to apprehend them. In this process, memory spaces and immaterial knowledge have been patrimonialized. We can cite as an example Brazil, where candomblé "terreiros" (space used by religious practices from African diaspora) have been listed as heritage sites, and processes of food production become patrimony, as the "bolo de rolo" (a rolled cake stuffed with guava), typical product of the State of Pernambuco, in the northeastern region of Brazil.

Facing this scenario in which the maintenance of memory becomes a responsibility of all, this falls into a moral imposition of keeping those memories, taking here the thought of Henri-Pierre Jeudy (2005), and considering everything that has been commented, would not it be appropriate to move on to the next topic to talk a little about the understandings on relationships between people and the elements of nature?

2. Interaction between people and elements of nature

The interaction between people and the elements of nature, as common sense suggests, can be said to have always existed. People, inserted in their cultural universe, designate the objects or elements of nature. The interrelation between the subject and the environment is formed by mental formulations of concepts which are fixed in the time and are maintained or transformed during the development of the history (Medeiros, 2009). Thus, from this interaction, marked by a chronology and a space, the knowledge of people about nature was forged.

Speaking about the origin of human knowledge, it is not possible to say when the production of knowledge began, nor who was the first person to give the decisive start. In fact, it can be said that the origin of knowledge has the age of man on Earth. However, the tradition of Western culture usually situates in the Classic period, in Greece, intermediate moments of what, at the same time, came to be called production of knowledge. Key historical personages in this process were Socrates, Plato, Aristotle, and Pythagoras, who initiated the structuring of an investigative methodology.

In this context in which we are talking about interaction and knowledge of people on the elements of nature, it becomes interesting to launch the questions of Bernardez (1981) which we reframe now: How do people relate to the environment? And, what are the types of positive and negative values that govern these relations throughout history? These queries are directly relationed to historical events that marked people's interaction with their living environments (Medeiros, 2009).

It is just through these interactions and the hability of people to produce culture and thus relate the natural elements to socially and culturally ways of understanding them, that humans have feelings and can give different meanings to the things (spaces, time, social aspects, and so on) (Bachelard, 1993). This way, with a creation of images and statements that remain as a stock of memory, people give to environments their forms of visibility, that are guided by a variety of filters, such as values, customs, political, ethical, and socioeconomic interests (Foucault, 1999).

It is worth saying that if we consider this interrelation between people and the elements of nature, we must have in mind that in this dynamic there is not only the cognitive but also the biologic aspect. Ecologic and evolutionary pressures and processes are essential components of the dynamics between humans and the environment. Another aspect that we must comment is that cognitive, ecologic and evolutionary processes suffer a qualitative and quantitative modulation in space and time (Medeiros, 2009).

The dating of the direct interrelation between the human species and its environment is already confirmed in the Neolithic age (or "polished stone age," which ends near 3000 BCE), when humans begin to dominate "the great arts of civilization: ceramics, weaving, agriculture and domestication of animals. [...] Each of these techniques supposes centuries of active and methodical observation, bold and controlled hypotheses, to be rejected or proven by means of tirelessly repeated experiences" (Levi-Strauss, 1962, p. 34). This aspect of curiosity of the humans ancestors, or this desire to "know for the pleasure of knowing," even because only "a small fraction [...] of observations and experiences [...] could give practical results immediately useable," that made the man of the Neolithic age the heir to an early scientific tradition (Levi-Strauss, 1962, p. 35). But, even before this period, the *Neanderthals* (Paleolithic era, approximately 200,000 years ago) had already established relations with the environment through processes of observation and experimentation producing a stock of conceptualizations, images and practices related to their spaces of life, including species and all elements of nature (Arvigo and Balick, 1998; Plotkin, 1991).

With respect to these ancient interactions and moving towards the relation between *Homo sapiens* and its natural surroundings, it was found that this interaction allowed human species "supply its organic, economic, and cultural needs, that is to say, daily necessities of food, drink, medicine, clothing, shelter, transport, etc., as well as the expression of imaginary and beliefs through rituals, legends, games, and arts" (Medeiros, 2009 p. 24). To illustrate this historical process of people's interrelations with the elements of nature, we will briefly highlight two fundamental aspects in the constitution of human life, namely, nutrition and therapeutics.

Starting with the question of food, it can be said that civilization would have been impossible without agriculture, without the cultivation of land.

To this essential activity to humanity, one has the domestication of animals susceptible to the raising of herds (Baker, 1968). Therefore, before the beginning of agriculture, the humans, as hunters and gatherers, nourished themselves through this activity.

During the Lower Paleolithic era, the humans could then ingest plant and animal species which contain sugars and vitamins, oils and starches as a source of carbohydrates, and most proteins came from meat. Considering the evolution of humans' nutrition habits, there was an increasing use of hunting weapons. Long after the development of skill for the manufacture of weapons, for the control of fire, for the construction of houses and the making of fur clothes from the animals that they hunted, humans started activities related to agriculture, already in the Mesolithic or Stone Age (13,000–9000 BCE). So, instead of having a way of life based on hunting and gathering, what belongs to a nomadic existence, most people began to stay in a determinate locality for a period, and because of that, communities were constituted. This new condition of staying in a place by human communities was the crucial transformation for the success of the development of agriculture (Baker, 1968).

It is interesting to observe that with the development of agriculture, people were selecting certain species for use in food, and today, although agriculture has improved the production of cultivated plants, there is a maintenance of this list of historically selected species. This way, it is possible to affirm that almost all the foods that integrate our daily nutrition are the same as those cultivated by farmers who started the agriculture (Baker, 1968).

Turning to medicinal plants and animals, these comprise a category of species that received great attention and importance because of the concern of the human species with the maintenance of their physical integrity from the beginnings of civilization. They are species of value for maintaining health or for curing diseases, for integrating religious ceremonies and practices linked to witchcraft, and more recently as sources of curare, reserpine, and other drugs of the 20th and 21st centuries.

Here, we will dwell a little more on certain aspects related to the interaction of people with the elements of nature, focusing specifically on these resources of therapeutic use.

2.1 Comments on the origins of the knowledge about medicinal virtues of the species

Whithin the ambit of medical research, a central interest is to know how people produce knowledge in this subject and which were the mechanisms to discover the therapeutic virtues of plant and animal species. In this context, we must ask ourselves what resources and techniques have been used by humans throughout evolution process to discover and select useful species to be employed as medicine. Let's look at some of the techniques that have been used by people to find out these medicinal species throughout human cognitive history.

2.1.1 Techniques for discovering useful species as medicine

2.1.1.1 Technique of nature observation

A common basic technique employed by humans in the origin of the process to discover useful species probably was the observation of the phenomena and characteristics of nature and the empirical experimentation of these resources (Arvigo and Balick, 1998). Through this technique, humans have been evaluating several species which have suggested the potentiality of their uses not only as medicine, but also in various aspects of human life, such as food, clothing and hunting weapons.

2.1.1.2 Technique of random attempt of hit and error

Another technique for selecting species used by people, that is also used in scientific researches has been the empirical use of species and the evaluation of the signs and symptoms after its use (Arvigo and Balick, 1998). The selection of species have been a step before these empirical use and evaluation process. So, the selection in this technique of random attempt of hit and error have been based on the quality of answers about the potencial of the use of species.

2.1.1.3 Technique of animal observation

The technique of animal observation is another basic manner developed by humans in order to discover and select species, and it is still used for the discovery of new chemical constituents of several species. With the observation of the behavior of domesticated animals or not, people have discovered the medicinal virtues of species. We can mention the classic example to illustrate this technique that is the exciting effect of coffee (*Coffea arabica* L.), discovered through the fable of Mullah Schadelich (Font Quer, 1967), a goat herder who observed the excitement of these animals after eating the fruits of this plant species. From this finding in the behavior of the goats, Mullah Schadelich prepared a drink with the fruits of the coffee to stay awake.

2.1.1.4 Technique of contact between cultures

Plants and animals can enter a medical system through a myriad of processes. For example, admission can occur through contact between cultures (Palmer, 2004). Thus, the goal of this technique of contact between cultures is to understand how interactions between humans from different origins, cultures and environments occur and promote exchanges of species between them.

2.1.1.5 Technique of expert's experimentation

This technique of expert's experimentation is centered on the figure of a specialist who conserves knowledge about the vitues of the species. So, the experimentation here is based on the actuation of specialists who test species for the treatment of diseases in their communities (Garro, 1986). From this expert's experimentation, the species that contain medicinal virtues entered into the local pharmacopeia and medical system (Phillips and Gentry, 1993).

Focusing on the factors that may be associated to the entry of a species into the local medical system from the experiment, two factors are pointed out in the literature, as follows.

2.1.2 Factors of availability and efficacy associated with the entry of a species into the local pharmacopoeia

Here, a common and usual question is always formulated. Why do some species enter into the local pharmacopeia and others not? We can answer this issue that there are certainly two important factors to be considered. One is the greater availability and accessibility of some species to people. The other, no less important than the first, and we can even say that is the motivating cause *par excelance,* is the efficacy of the therapeutic properties of these species. This way, the sum of these factors: availability in nature and accessibility by people plus therapeutic efficacy contribute together to the entry of some species into the local pharmacopeia. In short, relating to availability and accessibility, for Phillips and Gentry (1993), the resources available and accessible to people tend to present more experimentation. Relating to efficacy, Gottlieb et al. (1995) demonstrated phylogenetic terms in a list of medicinal plants for three indigenous ethnic groups of Amazon region (Brazil) which is concentrated in botanical families that are more evolutionarily derived and that the set of food plants is concentrated in the most basal families. This postulates that the selection of plants is not based on a randomly in botanical families. So, there is a group identified as being the overutilized families that englobes many family species with medicinal uses in face of the totality of family species available in that region. This can suggest differences in chemical composition of the species among the various families, indicating that the selection of medicinal plants is focused on therapeutic specificities. The criteria used to select species considered as having efficacy to treat conditions of health and desease which are named as the doctrine of the signs, theory of opposites, and the organoleptic properties, we will comment on subsequently.

2.1.3 Criteria for selecting species effective in the treatment of diseases and for the insertion of a new species in the local pharmacopoeia

2.1.3.1 Criteria: Signatura rerum (theory of signatures or doctrine of signs)

The doctrine of signs is attributed to the Swiss-German physician Paracelsus (1493-1541). The principle is based on the idea that everything created by nature shows through its own image (form) the virtue (property) that it brings in itself. This means: the image reveals its property. So, through the external morphology or other elements related to the species (color, smell, disposition of organs and tissues, habitat, and others) the signs and secrets of species can reveal their therapeutic activity. In this sense, there are some historical narratives that consider the theory of signatures a technique widely employed by people to discover and select species with medicinal effects. One of these historical narratives used as a classic example, it is the observation of the use of willow bark (*Salix alba* L.). Its use has spread all over the world becoming popular among peoples with the name of salicylic acid, an organic compost that is isolated from the bark of this species and functions as antimalarial, antipyretic, and against flu states (Font Quer, 1967).

2.1.3.2 Criteria: Theory of opposites (Hippocrates and Galen)

The theory of opposites is attributed to Hippocrates (460-370 BCE) and Galen (130-200 BCE) and it has been observed at different times in several human communities throughout Latin America (Waldstein, 2006). This theory belives in the healing by opposites, being characterized by the contrary aphorism *contrariis curantur* (Font Quer, 1967). It has a great importance because it have determinated the origin of the procedure of allopathic medicine. Its healing process is opposed to the cure by the similar, which represents the main concept of homeopathy procedures. We can observe the application of this theory in the selection of medicinal species considering the hot—cold system. This is a system in which diseases are identified as humoral imbalance causing diseases that, if onsidered hot, are treated with species considered cold and, if considered cold, are treated by hot species. The finality of this system is to restore the humoral balance. So, the selection of species in hot-cold system is understood as being a reflection of how people feel and give meaning to diseases according to their culture values (Kleinman, 1978).

2.1.3.3 Criteria: Organoleptic properties

People through the taste and smell (humans' senses) made choice of medicinal species considering their organoleptic characteristics (Hinton et al., 2008; Shepard, 2004), and their influence in the selection of species can be understood in three strands: mnemonic, clue, and determinant. Starting from the mnemonic role, this updates knowledge and sensations from the memory of the past about some particularities of the species as well as their associations to their therapeutic efficacy. This process will be continuous and these traces of memory will remain and will be remembered for generations to come. The other role is the clue. The medical systems have a constant modulation. In reason of that, the experimentation and selection of species are guided by the intrinsic necessity of adaptation to new conditions of health and disease. The clue lays on the search for species that have certain characteristics that are already fixed in people's memory and will be of help (Casagrande, 2000; Peroni and Torry, 2007). And, finally, the determining role in the selection of medicinal species refers to the influence in the choice of species that carries certain properties. This influence would be caused by predisposed caractheristic of humans related to the recognition of certain tastes. The direct consequence of the determining role is the association of species with properties indicated by the senses that point out their medicinal quality. There are two models that deal with the predisposition of the human species to organoleptic properties: human chemical ecology model (Johns, 1990) and sensory ecology (Shepard, 2004).

In the human chemical ecology model, Johns (1990) proposes a model of coevolution between humans' perception of taste and smell and the chemical compounds inherent in other species, specially regarding the compounds that integrated their alimentation. In this perspective, people have the potenciality to recognize certain classes of chemical compounds in the environment, discerning what is good or toxic from species. When working out his ideas, the author believes that humans in their historical evolution developed their capacity to know what it would be favorable or not for them to eat. In this line of thought, Johns explains that through the effects of chemical compounds in food, human medicine originated from an intelligent response given by people in the face of the adaptations necessary to survive in a better way. In addition, this coevolutive condition occurs through biological adaptations and cultural traces, which are related to behavioral and physiological adaptations (biological) and to the ability to communicate, transmit and learn konwledge and practices about nutrients and toxicity of the species (cultural). It is also important to mention that herbivores and the domestication of species by humans contribute to selection processes about the presence of chemical compounds in plants.

By talking about the model of sensory ecology proposed by Shepard (2004), in reality, this proposition dialogues with the model commented above, when Shepard gives weigth to the idea that the sensations are a biocultural product resulted by the adaptation of humans to the recognition of classes of chemical compounds of the environment. Beyond this biological aspect, the author also considers that the manifestation of sensory properties applied to the use of species by humans suffer influences of individual experiences and culture (Fabrega, 1997; Shepard, 2004).

Until now, we observed that the selection of medicinal species by humans may be associated with several factors. Following this idea, Casagrande (2000) develops the concept of medicinal prototypes. In this sense, he considers several elements, such as taste, smell, and color which may influence the selection of medicinal species. This way, species evaluated as medicinal prototypes form a small number of species existing in a locality (Casagrande, 2000). The information about this process of recognizing the environment may be related to the memory. Thus, what happens is that any information stored mentally is related to some elements elected as being essential for survival and important to be transmitted to subsequent generations.

Other forms of discovery of the therapeutic effects of species are also found in innumerable practices among some human groups, for example, through the practice of sacred rituals, or the use of amulets, and even to heal through prayers that often use benzene plants. These manifestations have contributed to the selection and incorporation of natural resources used for medicinal finalities, and also for other purposes. These evidences point the relevance not only of social actors considered specialists from the earliest times, such as alchemists, healers, and shamans, but also the knowledge of nature disseminated by the population.

After these considerations about the process of discovery of biodiversity by the human species over time, let us review some examples of documentary evidence that constitute records kept to the present day that reveal different aspects of the interaction between people and the elements of nature.

3. Documentary sources: past evidence about the history of people with nature

Let us now take a look at some documents, historical sources, which will point us to a reflection on the insertion of plants and animals into the constitution of human society. They are reports of chroniclers, religious and naturalists on European, Eastern, and American nature. By means of these data, which have as their basis the intimate relationship that is established between people and the plant and animal universe over time, we have before our eyes the revelation of what historical ethnobiology represents.

The history of plant and animal species predates the human species at least 470 milion years ago. Through evolutionary processes, when our history is compared to those of other species, we observe that our timeline referers to a not too distant time scale. Since then, plants and animals have been playing a primordial and dynamic role in shaping our history.

These elements of nature are our source of food, medicine, fuel, shelter, clothing, arrow poisons such as hunting tools, body paintings, rituals, etc. All these uses appear in the early stages of human prehistory. That is, they were determinants and influenced the history of our life on the planet and continue to be essential for our survival, for our development (Laws, 2013). Some uses of plant and animal species reflect more the taste, the aesthetic character, of people than their need for survival. For this reason, these uses may be considered minor compared to the use of species for food and medicine (Balick and Cox, 1996).

Taking vegetable species as an example, we can say that they guided the path of human civilization. To illustrate this line of thought, we can briefly mention the importance of Marco Polo's description during the 13th century of an island where there were many spices species, such as pepper (*Piper* sp.) and nutmeg (*Myristica fragrans* Houtt.). The written document of Marco Polo generated great interest and ended up being the propeller of the intense navigations by Europeans that were given in search of the spices, which culminated in the circumnavigation by Magellan, when occured discovery of America. Like spices, so many other species of commercial interest, such as rubber and quinine, have also influenced the history of nations around the world (Balick and Cox, 1996).

As the production of people's material culture is particularly rooted in plants, let us then take some of these historical accounts as examples of sources that relate to the past use of plants by many human societies, through evidence with graphic or written representations of their interactions with plants.

3.1 General references

3.1.1 East and Europe

Throughout history, ancient civilizations such as those of the Egyptians, Sumerians, Assyrians, Babylonians, Greeks, Persians, Chinese, and Indians had great knowledge about the potentialities of nature's species. The heritage of knowledge of these civilizations has been preserved in registers that where transcription of this know-how is kept as a legacy. Given the importance of these testimonies, we will go through each of these civilizations a little, seeing examples of documents that contain this knowledge.

Firstly, we will consider the Ebers Papyrus and Smith Papyrus, both from Egyptian origin, dating back to 1500 BCE and they contain the transcription of more than 700 formulas from the Egyptian pharmacopeia. During this historical moment there was an intense commerce between Egypt and the cities of the Kingdom of Sabá, located southwest of the Arabian desert (present-day region of the Republic of Yemen). Through this trade many drugs composed by plant species arrived in Egypt for consumption (Pelt, 2003). Studies continue to date from the time they were discovered, more than 100 years ago (Anderson, 1977; Baker, 1968; Pires, 1984). The Ebers Papyrus was the first known Egyptian medical treatise, deciphered by the German Egyptologist Yorg Ebers in 1873. This treatise is dedicated to the treatment of internal diseases, and also contains indications on the composition of the medicines to be used, many of them being aromatic plants. From this document, the world became aware of the description of medicinal plant species from this Egyptian pharmacopeia, such as a sedative and antispasmodic species so-called poppy (*Papaver somniferum* L.), that is the source of opium, morphine, codeine, and papaverine.

Among the Sumerians, the transcription of their knowledge and practices were preserved through cuneiform writing, in clay plaques. Of these plates, several recipes have been translated. In one of them, there is a reference to the collection and ritual use of, as in the Egyptian papyrus, the poppy. Among the Assyrians, their knowledge was maintained through the production of an herbarium which contained several species used as drugs, such as mandrake (*Mandragora officinarum* L.), juniper (*Juncus* sp.), and thyme (*Thymus vulgaris* L.) (Almeida, 2003; Cotton, 1996).

From the region of Asia Minor, Herodotus (about 484–25 BCE) recorded traces of memory related to the Greco-Persian wars that occurred in the course of the 5th century BC (490 and 480–79 BCE). *The Histories*, as was called Herodotus' legacy, is composed by the transcription of stories and history compiled through interviews made in the period when he traveled in this region (Schwartz, 2008). Herodotus was the precursor of ethnographic studies and brings in his narrative information like that about the process of pollination by Babylonians, geography, and zoology (Herodotus, 2014).

Later in history, Theophrastus (372–287 BCE), disciple of Aristotle (384–22 BCE), transcribes observations made by Alexander the Great (356–23 BCE) and gathered and registered much information on plants (Balick and Cox, 1996). After Alexander's death in 323 BCE and the collapse of his empire, Alexandria became the scientific center of the world for approximately three centuries.

From the East, specifically from China, comes the oldest record named *Pen-ts'ao* (or Book of Herbs, published about 1578 BCE). This register originated from Emperor Shen-Nung's own oral narrative, who was also an herbalist and lived during the 2800 BCE, is composed by the description of 365 herbal remedies (Balick and Cox, 1996; Cotton, 1996; Occhioni et al., 1979).

During the so-called classical civilizations, plant drugs began to be systematically recorded. With the influence of Greek medicine and the Egyptian school, we refer to Hippocrates (460–370 BCE). In the writings left by Plato (428-347 BCE) and Aristotle, Hippocrates is referred as being the prototype of physician and the very personification of medicine (Carvalho, 2002). Hippocratic medicine assumed that the causality of disease would not be in divine or supernatural belief. He developed a system of empirical medicine that influenced the formation of the Western medicine, that is, there were an approach to diagnosis and treatment based on clinical experience (Hippocrates, 1555; Nepomuceno, 2003).

In Italy, in Rome, Galen of Pergamus (130–200 BCE) developed the method of polypharmacy, which comprises a manner for handling powders and extracts. This method had a great insertion in the pharmaceutical practices and medical studies and it was transmitted to academics for hundreds of years (Galeno, 1550; Nepomuceno, 2003). After the creation of this method, the pharmaceutical preparations composed of more than two drugs have come to be identified as galenic therapy.

Reaching the Christian era, between the years 33 and 79, the roman naturalist Pliny (23-79 AD) wrote *Historia Naturalis*, an encyclopedia composed of 37 volumes, being eight of them about descriptions of plant species that were in the practice of Roman uses (Pliny, 1855).

Another special heritage was the work of *De Materia Medica* written by the Greek physician Pedanius Dioscorides (40-90 AD) in 78 AD (Dioscorides, 1554). He was born approximately in 64 AD and traveled extensively, especially as a surgeon serving the army of Emperor Nero (37-68 AD). His work, written, illustrated, and conceived in a "herbal" format, was for more than 1500 years a reference for physicians and apothecaries. This register comprises 600 descriptions of plant species and their medicinal virtues. For the West, *De Materia Medica* is considered the first treatise of phytotherapy (Almeida, 2003; Luccioni, 1999; Nepomuceno, 2003). This manuscript was copied and recopied for the next 16 centuries after its publication, having become an icon for the works dedicated to botany. Just as the theme of this document was medicinal plants, the other writings already mentioned, as well as others that we will mention, have also dedicated themselves to the same subject. This is due to studies about the therapeutic actions of medicinal plant species have influenced the science of botany. This is the reason why in the history of botany the earliest botanists are termed "herbalists, an allusion to the registers on the medicinal use of species. This fact makes it difficult to systematize the records produced in this first period of botany, as there were many essentially botanical productions and others essentially on the medicinal use of plants (Lorenzi and Matos, 2008).

Considering the period after the fall of the Roman Empire and during the Middle Ages (between the 5th and 15th centuries) we can observe that there was a transformation in the European continent, including scientific contributions. In the 5th century, as a result of the continuous wars for the conquest of its territory, Italy suffered a rapid decline: commerce and manufactures were extinguished, land was no longer cultivated, plague and famine devastated the population, and the population declined consubstantially. However, monasteries and convents became major centers of culture. This is because these religious spaces assumed the responsibility of preserving ancient records, such as the alchemy and medicine treatises in which were reported the use of natural species and other elements of nature (Nepomuceno, 2003). In this context, Saint Benedict of Nursia (480—547) helps the religious to be zealous for safeguarding the classical culture by founding the Benedictine religious order. Then, around 529, St. Benedict founded the monastery of the Benedictine Order, the Monastery of Monte Cassino, and writes his *Rule* of life to be followed by the religious in all Benedictine monasteries (Bento, 2003; São Bento Monastery of Rio de Janeiro, 2004). Parallel to the historical-social events, it is important to think about the meticulous work of the Benedictine copyist monks. By doing this precious work, they helped that the knowledge about medicinal species, literature and other records left by the Greeks, Romans, and Arabs, were preserved and have reached the present day (Brasil and Galdós, 2004; Duniau, 2003).

In the 10th century, the Arab world emerged with great scientific activity. From there arises the "Arab medicine", a medicine influenced by Greek and Indian principles, that spread throughout Europe. The iconic figure of this medicine was personified in the physician Avicenna (ca. 980–1037), who is appointed as the developer of aromatherapy: flowers as a therapy for heart disease (Almeida, 2003; Nepomuceno, 2003). Avicenna's *Liber canonis medicine* brings centuries of classical medical knowledge (Avicena, 1527). Across the Iberian Peninsula, Arab knowledge gained the whole of Europe. Many drugs, new to the era, were introduced in European therapeutics, such as cinnamon (*Cinnamomum verum* J. Presl), lemon (*Citrus* sp.), and camphor (*Cinnamomum camphora* (L.) J. Presl).

From Germany, we can remember St. Hildegarde of Bingen (1098–1182). She was thorough in describing the properties of German plants and animals. Superior of a Benedictine monastery in Germany, Hildegarde made exact diagnoses and effective formulations (Giannini, 2004). Her two medical texts *Liber simplicis medicinae (Physica)* and *Liber compositae medicinae (Causae et curae)*, where the first was adopted as a reading even by the University of Montpellier in France, denote the concern of hildegardian medicine in caring for the patient, recovering, maintaining, and protecting the natural health of man (Brasil and Galdós, 2004; Giannini, 2004; Pernoud, 1996).

Following the "herbal" format, many records were produced during the 15th century in Europe by the development of the press. Many of these documents were very similar to the original "herbal" written by Dioscorides. This aspect has changed in the 16th century, when the "herbal" has improved the quality of production. Among these "herbals" considered of quality, we can cite the *Institutionum medicinae libri quinque...*, published in 1594 by the German physician and botanist Leonhart Fuchs (1501-66) and the *Historia plantarum* by other German physician and botanist Valerius Cordus (1515-44), published in Prussia in 1561 (Fuchs, 1594; Cordus, 1561). In these works, the authors observed the previous botanical studies and the nature in order to register descriptions, illustrations and several information regarding the habitats, uses, management, and there was also a narrative permeated by myths and legends involving species such as the mandrake.

To finalize these brief words about some of the registers that represent the ancient memory from East and Europe, we choose to talk about an heritage that is considered even "[...] one of the most important books ever published on plants ever published in the English language" in the words of Balick and Cox (1996, p. 14). This last trace of memory that we would like to comment on here is the register entitled *The Herball, or Generall Historie of Plantes*, written by the British barber-surgeon John Gerard (1545-1612) (Gerard, 1633). Published originally in 1597, this work soon became a source of consultation for practitioners of healing arts, since its content translated the knowledge accumulated since the time of Dioscorides on the medicinal potentialities of various plant species, including those discovered in the Americas.

3.1.2 Americas

In the Americas, geographic discoveries at the end of the 15th century, with the opening of the sea routes to the Indies and to America, and explorations for the recognition of the economic potential of plants, animals, minerals, domain, and expansion of colonies were a current practice.

During centuries of this practice of colonization occured a cultural miscegenation, which has resulted in a so-called "universal" process by which social groups identified as colonized and colonizers have absorbed or repudiated new conceptualizations. With the occurrence of these new interrelations, there was the possibility of exchanging knowledge that was previously private, exclusive or proper to a society, among the social groups involved (Medeiros, 2009).

Throughout America, native societies have undergone significant changes, whether in the aspect of their territorial occupations or their cultures, and thus have undergone complex fission and fusion processes, even among ethnically differentiated groups (Wüst, 1998). These new ways of being in the environment were a consequence of the direct and indirect contact with the different societies promoted by the European colonization of the new territory.

However, we cannot agree that it was without social conflict that this whole process of colonization and, as a result, cultural miscegenation was established. These cultures and their environmental manifestations have been deconstructed and reconstructed, expropriated and reappropriated throughout America by various optics between peoples and elites. Thus, the appropriation and use of environmental resources was guided by political-economic processes and constituted by diverse processes of economic rationalities, based on different perceptions, meannings, feelings and "needs" of nature in the face of a wide possibility of social actors (Cabral, 2008).

The new conditions of social structure and environment forged the appearance of new needs on the part of the colonized as well as the colonizers. Thus, the yearnings of the colonizers and the colonized determined the activities, relations, positioning, behaviors, etc., related to the environment. Among the things that the settlers had in mind, one of them was the domination and the economic gain through the exploitation of the natural resources of the territories conquered from the Discoveries.

In this scenario of conquest of new territories and the discovery of new cultures and biodiversity by European colonizers, botanic gardens and curiosities offices emerged. In the 15th and 16th centuries, these spaces began to systematize knowledge about the uses of species and to collect exotic "things" discovered during travels across the American continent. This was a practice that aimed to ensure dominance of the potential arising from the discoveries of these new territories. This process of collecting and maintaining a set of "things" (objects) that reveals ideas on the same theme resulted in the development of the areas formed by the collection, which are art history and natural history areas (Janeira, 2006). At that time in Europe, this tendency to preserve an heritage of memory through objects contemplated both the artistic expression of the Renaissance and the fashion of gathering exotic specimens from the new territories and the nature of the Americas.

From the perspective of the colonized, the knowledge and practices of exotic species were incorporated to their place of origin, in such cases as the occurance of new diseases caused by colonizer—colonized contact. Or when, in the process of colonization, external agents (social factors) came into contact with the local population and from this contact, there was the spread of diseases and, as a consequence, a demographic collapse and a change in the social structure. In this process, introduced plants and animals and wild plants and animals are used for medicinal purposes.

Thus, we can say that a good part of the so-called discovery was the transfer of knowledge from the native Amerindian cultures and species that the American continent possessed to the societies, cultures, and economies of the European man, from the Americas to the Old World, for example, tobacco (*Nicotiana* spp.), corn (*Zea mays* L.), and cotton (*Gossypium* spp.).

In this way, we can postulate that botany and zoology are generally related to an epistemology, including theories and methods, which included devotion to science but, at the same time, it could occur an adaptation to political, economic and cultural scenarios and to the wishes of the dominant social actors (Klanovicz, 2002).

In the course of history, botany and ethnobotany, as well as zoology and ethnozoology, were apprehended in the same discipline. This is a universalist view that considers that throughout the course of human history the authors expressed their point of view through the records of information on the utilitarian knowledge and practices of nature species in view of new natural products which added economic gains. So, the action of figures such as explorers and naturalists were permeated by both desires to investigate and systematize the species and to develop the potential of the economic gain through the exploration of these natural resources. In this sense, ethnobotany and ethnozoology were closely related to botanical and zoologic exploration, and the history of these sciences could be understood as having a parallel to the development of systematics (Davis, 1995), if we consider a universalist point of view.

The documentary heritage left by the chroniclers is also very striking in their detailed account of the civilizations of the Americas. From the 16th century the American documents sought to discuss the nature and wisdom of the peoples of overseas. Thus came records that spoke primarily of hallucinogenic, medicinal, and food plants. Among these treaties, a few examples may be cited.

One of them is the *Historia General de las Cosas de la Nueva España* (or *Florentine Codex*), a book dating from 1538—60 and written by the Spanish Frei Bernardino de Sahagún (1499-1590), who arrived in Mexico in 1529 (de Sahagún, 1989). According to Taube (1993), this writing constitutes the most detailed chronicle of pre-Hispanic culture, where more than 100 vegetables, especially hallucinogens, are described. Sahagún worked in the *Colegio de la Santa Cruz de Tlatelolco* (Mexico) and wrote the *Florentine Codex* with the help of Nahua students from the same school and also older people from various villages in central Mexico (Biblioteca Medicea Laurenziana, 2007).

Another example is the *Libellus of Medicinalibus Indorum Herbis* (or *Codex Badianus*), dated 1552 and elaborated in Mexico by the indigenous physician Martín de la Cruz (ca. 1510-55), from Santiago de Tlatelolco and associated with the *Colegio de la Santa Cruz de Tlatelolco* (Guerra, 1952). The author collected information on various aspects of the Aztecs and described the medicinal properties of 200 plants used by this civilization. Written in the *Nahuatl* indigenous language, the work was translated into Latin by Juan Badiano (1484-after -1552), who was also a student of the *Colegio de la Santa Cruz de Tlatelolco.*

We can also mention the Spanish chronicler and Dominican friar Bartolomeu de las Casas (1474-1566), who participated in the second *voyage* of the Italian Christopher Columbus (1451-1506) to the Americas. He belonged to the school of Salamanca (Spain) where he was engaged in legal-theological debates and professed a natural philosophy. Las Casas applied this theoretical vision in his documental production about what he observed in the New World, giving a divine trace to this new environment and its peoples. He wrote the *Apologetics Summarized History* (de las Casas, 1967), making a vast ethnographic contribution and going in defense of the Indians rights, and so, considering them as human beings. In his work he also presents a thorough description of the nature, including botanical observations pointing out the relation between the images of the species with their therapeutic virtues, where the images are reflective signs of properties of the species, thus revealing strong influence of the doctrine of the signatures of Paracelsus in his discourse.

Among the first registers in Europe to the use of medicinal plants and substances from overseas lands are the writings of two Iberians: Nicolás Monardes (1493-1588) and Juan Fragoso (1530-97), entitled *Segunda parte del libro, de las cosas que se traen de nuestras Índias Occidentales que sirven al uso de medicina [...] e Discursos de las cosas aromáticas, árboles y frutales, y de otras muchas medicinas simples que se traen de la India y Oriental y sirven al uso de la medicina,* published in Spain in 1571 and 1572, respectively (Fragoso, 1572; Monardes, 1571). Various species with therapeutic virtues are presented in the works of Monardes and Fragoso, and the cinchona (*Cinchona* spp.) appears as a medicine indicated for fevers. It was thanks to the insertion of secular indigenous knowledge about the use of cinchona in European medical practice, patients suffering from fevers have been cured without the usual practices of Western medicine such as bleeding, purging, and vomitory, practices based on the humoral theory, widely accepted at that time (Crespo, 1995; 1997). The scientific contribution brought by Monardes and Fragoso was relevant and it contributed to a new conception of health and disease within academic medicine.

The cinchona can be taken as a forceful example to understand the doctrine of signatures. In the account of Father Antonio de la Calancha (1584-1654), *Crônica moralizada del orden de San Agustín em el Perú,* there is a description of the incident with a Spanish soldier (de la Calancha, 1638). This man was healed from malaria after had drunk water from a puddle in the rainforest. The curious thing is that near this puddle there was a tree that indigenous attested to have healing power. They told about curing sick animals when they drank water from some ponds that had this specific tree on the banks. The priests, when they came into contact with this indigenous belief, diluted the bark in the water and verified its bitter taste. From this observation, the religious were related to the precepts of the doctrine of signatures. This way, the bitter taste of the drink came to indicate the existence of some use of this species for man.

An equally important account of the discovery of the medicinal use of cinchona was published in the booklet published in 1663, *Anastasis Corticis Perviae,* by Sebastiano Bado (1640-1676) from Italy (Bado, 1663). Bado comments that the wife of the Viceroy of Peru Fernándes de Cabrera (1589-1647), Count of Chinchón, suffering from malaria was also healed with the bark of these same miraculous tree. The author states that the countess due to this fact, when she went back to Spain, she transported a great quantity of barks of this healing tree. It is noted that in honor to the countess of Chinchón, vice-queen of Peru, cinchona had its botanical classification as *Cinchona officinalis* L. Soon after, the Jesuits released this medicine in Europe and were responsible for making this bark known and used by populations.

Let's talk in more detail about some documentary evidence produced throughout history produced in a megadiverse country of the American continent, which is Brazil. The documents produced during the historical process and that refer us to the history of exploration and progressive discovery of the Brazilian biodiversity can be correlated to the periods of the history of botany.

Therefore, following a periodization inspired by Ferri's *History of Botany* in Brazil (1980), it is possible to classify as an initial period the one dedicated to the production of chroniclers and missionaries, whose deadlines would be between the 16th century and the mid-17th century. On the flora and fauna of Brazil, one can say that the first documents produced were the treatises and letters addressed to the king, such as the letter of Pêro Vaz de Caminha (1450-1500) to King D. Manuel I of Portugal and Algarves (1469-1521) in 1500 (de Caminha, 1500). Thus, the first look at the Brazilian flora and fauna was recorded by chroniclers, colonizers, and missionaries, as in the work entitled *História dos Animais e Árvores do Maranhão*, by Frei Cristóvão de Lisboa (1583—1652) (Alcides, 2005; Walter, 2000).

From the colonization (second half of the 16th century) came the Jesuit medicine and the official medicine brought by physicists from European universities, being Jorge Valadares (-1557) considered the first physicist in Brazil (Scliar, 2005). The so-called Jesuit medicine reflects the medical experience of this period in which the religious of the Order of the Jesuits acted as doctors, nurses, and apothecaries of both the indigenous and the colonizers, and their infirmaries were the first hospitals of the colonial period (Duniau, 2003; Santos Filho, 2005). Among the actors of this period, the Jesuit José de Anchieta (1534-97), who arrived in Brazil in 1553 with the second General-Governor Duarte da Costa (1505-60), is pointed as the best physician figure of the16th century in Brazil (Santos Filho, 2005).

The Jesuits already in Brazil, in 1549, were dedicated to education and prioritized the knowledge according to a scholastic philosophy. Working for teaching they implemented the first schools in Brazilian territories. In favor of the unification of the indigenous languages they made semantic and phonetic junctions of them, resulting in the *Tupi* language, thus leading to a single common language. Hospitals and pharmacies were also the proposals in which they were also engaged (Cabral, 2016). Their pharmacies were installed in some regions of Brazil, in the present states of Bahia, Pernambuco (Olinda and Recife), Maranhão, Rio de Janeiro, and São Paulo (Fernandes, 2004; History of Brazilian Pharmacy, 2005a).

By linking knowledge of European medicine to indigenous medical practice, the Jesuits presented indications and effects of native species that were eventually incorporated into the world pharmacopoeia (Santos Filho, 2005). For the manufacture of medicines, the Jesuits counted on raw materials from Europe, but access to these products was becoming difficult to obtain and the Jesuits started to make medicines with local Brazilian species.

In the Jesuit School of Bahia, along the time, a preparation that incorporated components of the Brazilian flora in the preparation of the famous *Triaga Brasilica* was developed. This triaga is a pharmaceutical preparation that is an electuary including the mixture of extracts of several plants, an inheritance of ancient Rome (Duniau, 2003; Carneiro, 2005; Pereira et al., 1996). This incorporation of Brazilian species into the triaga formula were made by the Jesuits through the knowledge of the indigenous peoples and they also established a comparision between native and the original European elements of the formulations.

The *Triaga Brasilica* was used in febrile diseases, colic, applied against all types of animal poisonings, in the case of venomous animal bites, abdominal pains, intestinal worms, diarrhea, "head problems", epilepsy, apoplexy and melancholy, plague and other epidemic diseases, malaria, smallpox, measles, "hysteria and diseases particular for women" (Cabral, 2016; History of the Pharmaceutical Industry in Brazil, 2005b).

This recipe so precisous motivated the Marquis of Pombal (1699-1782) to issue the order to banish the Jesuits from Portugal and the colonies of Brazil. This political act, in 1579, would allow the possession of this recipe by the Portuguese Crown. But, this act by Marquis of Pombal was uncessfull and the recipe was not reached. Only in 1766, it was found again in a Jesuit archive in Roma, and finally revealed through the book *Artes e Ofícios dos Jesuítas no Brasil (1549-1760)* published by the Jesuit Serafim Leite (1890-1969), in 1953 (Leite, 1953).

As a second classificatory period in the history of documentary production in Brazil, there is the one of Dutch Brazil (1637-44), which occurred during the stay of the Dutch in the northeast of the country. It was characterized by a period of scientific activities carried out by the naturalists brought by Johann Moritz von Nassau-Siegen, Count of Nassau (1604-79), but that did not leave an institutional scientific heritage in Brazil.

Among the doctors in Brazil in the 17th century, the name of the Dutch doctor Wilhelm Piso (1611—78) stands out in Brazil during the Dutch stay in the Brazilian Northeast under the rule of the Count of Nassau (Santos Filho, 2005; Nogueira, 2000). Together with Georg Marggraf (1610—44), a German naturalist who also participated in the Dutch delegation, he explored the interior of Pernambuco, Paraíba, and Rio Grande do Norte, collecting specimens of the flora and fauna of this region (Nogueira, 2000). Piso and Marggraf had their experiences registered in works. Piso wrote the *De Medicinae Brasiliensi* and Marggraf the *Historia rerum naturalium Brasiliae*. These two registers were grouped in a single publication and printed in 1648 in Amsterdam with the title *Historia Naturalis Brasiliae* composed of 12 books (Piso and Marggraf, 1648) (Fig. 3.1). *Historia Naturalis Brasiliae* is one of the first major editions dedicated to the Brazilian biodiversity and peoples in the colonial period (Alcides, 2005; Cunha, 2004;

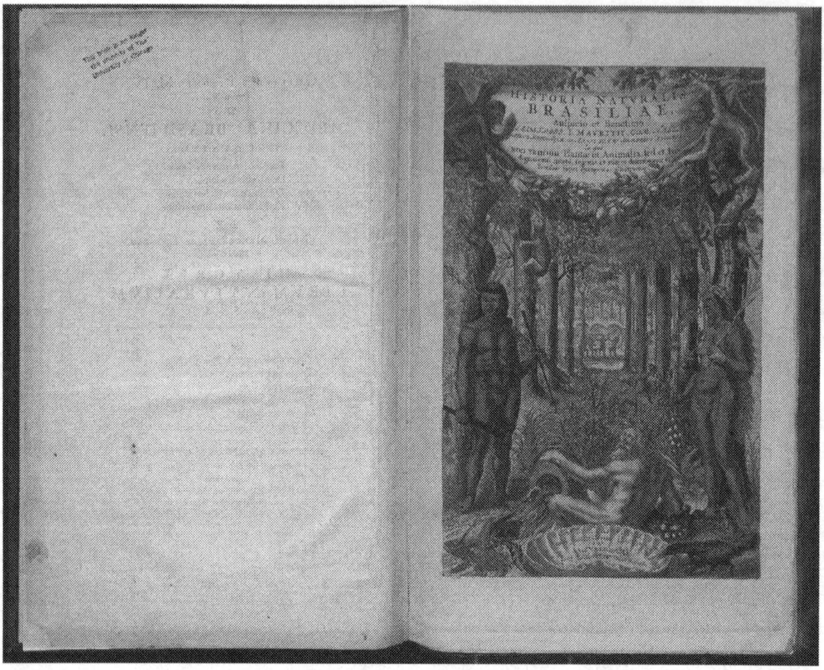

Figure 3.1 Front page of *Historia Naturalis Brasiliae*, hand colored woodcut, paper, 41×27,5 cm (open), belongs to José Antônio Gonsalves de Mello Library collection of Ricardo Brennand Institute, Recife, Brazil. *From Piso and Marggraf (1648).*

Marques, 1999; Nogueira, 2000; van den Berg, 1993). In their works, Piso and Marggraf sought information on some plants studied in one of the main sources on new medicinal species found outside Europe, consulting the work of the Portuguese physician Garcia de Orta (1501-68), *Colóquios dos Simples e Drogas da India*, published in 1563, where this author in his travels in the interior of India collects and describes new species, which translate a new medical matter learned (Orta, 1872).

The third period corresponds to the awakening of Brazilians, from the second half of the 18th century, with the departure of children from the elite of Brazilian society to the University of Coimbra. The formation of human resources to meet the economic interests of the Portuguese Crown marked this epoch. Some historical characters may be mentioned for the period.

Manuel Arruda da Câmara (1752—1811), traveling through the hinterland of Pernambuco to Piauí (present states in notheastern of Brazil) in 1794, gathered information on the use of the natural resources and expressed his concern with the extinction of native species throught registers he made.

He also influenced the Prince-Regent with his ideas presented in 1810, in his work *Discurso sobre a utilidade da instituição de jardins nas principais províncias do Brasil*, and in the next year, D. João promoted the establishment of botanical gardens in the northeast (Pernambuco and Bahia) and southeast (Minas Gerais and São Paulo) of Brazil (Nogueira, 2000; de Mello, 1982; Paula, 1998).

Also noteworthy are the names of some naturalists, such as Alexandre Rodrigues Ferreira (1756–1815), who was indicated by Domenico Vandelli and appointed in 1778 by the Portuguese government, at the request of Queen D. Maria I (1734-1816), to explore close to a scientific commission the natural wealth of the Captaincies of Grão-Pará, Rio Negro, Mato Grosso and Cuiabá, and he is also author of *Viagem Filosófica* and other unpublished works; José Mariano da Conceição Vellozo (1742–1811), author of *Flora Fluminensis* in collaboration with several specialists, who was one of the first to report information about our medicinal plants, followed by Francisco Cysneiros Freire Allemão (1797–1874), naturalist of the National Museum of Rio de Janeiro and professor of the Medical School; and Frei Leandro do Sacramento (1770–1829), appointed director of the Royal Botanic Garden in 1824 and first professor of Botany of the Medical-Surgical Academy of Rio de Janeiro (Ferreira, 1971, 1972, 1974; Pires, 1984; Nogueira, 2000; Vellozo, 1825, 1827, 1881).

The fourth moment was characterized by the presence of naturalistic travelers and the scientific institutionalization in Brazil, which occurred from the19th century until the mid-20th century. As a consequence of the biologic wealth existing in Brazil, the institutionalization of science occurred through the natural sciences. Among the foreign researchers dedicated to the study of Brazilian biodiversity are Sellow (1789–1831); Auguste de Saint-Hilaire (1779–1853); Riedel (1790–1861); Phillipe von Martius (1794–1868); and Richard Spruce (1817–93) (von Martius et al., 1840; Rodrigues, 1992; von Martius, 1979; Nogueira, 2000; Pires, 1984). These men of science dedicated themselves to learning and collecting specimens and produced an heritage full of knowledge about species and local customs, in some cases, with knowledge and practices information.

Finally, there was the contemporary period, from 1950, when there was the beginning of the creation of development agencies in Brazil, enabling the development, strengthening, and expansion of scientific research in their graduate programs and research centers (Fernandes, 2004). Lorenzi and Matos (2008) argue that in the 20th century most of the published publications consisted of compilations of works in their regional majority. It is worth mentioning the gigantic compilation work carried out by the Portuguese naturalist Manoel Pio Corrêa (1874-1934), *Dicionário das Plantas Úteis do Brasil e das Exóticas Cultivadas*, a collection of six volumes, of national character, launched from 1926 to 1975 (Corrêa, 1984).

References

Abreu, R., 2008. Patrimônios etnográficos e museus: uma visão antropológica. In: Abreu, R. (Ed.), E o patrimônio? PPG em Memória Social da UNIRIO, pp. 33–57. Rio de Janeiro.

Alcides, M.M., 2005. Historia Naturalis Brasiliae: Um estudo do registro botânico holandês seiscentista. Masters Thesis. Universidade Federal de Alagoas, Alagoas, Brasil.

Almeida, M.Z., 2003. Plantas Medicinais, second ed. EDUFBA, Salvador.

Anderson, F.J., 1977. An Illustrated History of the Herbals. Columbia University Press, New York.

Anon, 1772. Estatutos da Universidade de Coimbra, third volume. Universidade de Coimbra, Coimbra, pp. 266–267.

Anon, 2005a. História da Farmácia Brasileira. Available from: http://www.terradeos.com.br/ervas/farmacia.html.

Anon, 2005b. História da Indústria Farmacêutica no Brasil. Available from: sincamesp.com.br.

Arvigo, R., Balick, M., 1998. Rainforest Remedies: One Hundred Healing Herbs of Belize. Lotus Press, Twin Lakes.

Avicena, 1527. Liber canonis medicine. In edibus Luce Antonii Junta, Venetiis.

Bachelard, G., 1993. A poética do espaço. Martins Fontes, São Paulo.

Bado, S., 1663. Anastasis corticis peruviae; seu chinae chinae defensio, Sebastiani Badi ... Contra ventilationes Joannis Jacobi Chifletii, gemitusque Vopisci Fortunati Plempii ... Opus in tres libros distinctum, & in eis documenta medicinae & philosophie – Phlebotomiae necessitas ... in variolis, morbillis, exanthematis, etiam apparentibus. Petrus Joannes Calenzani, Genoa.

Baker, H.G., 1968. Las plantas y la civilization. Herrero Hermanos Sucesores S.A., Ciudad de México.

Balick, M.J., Cox, P.A., 1996. Plants, People and Culture: The Science of Ethnobotany. ScientificAmerican Library, New York.

Bento, S., 2003. A regra de São Bento. Latim – Português, third ed. Lumen Christi, Rio de Janeiro.

Bernardez, G., 1981. Ecologia y paisaje. Hermann Blume, Madrid.

Biblioteca Medicea Laurenziana, 2007. The World of the Aztecs in the Florentine Codex. Mandragora, Florença.

Brasil, A.F., Galdós, J.V., 2004. Atlas das plantas medicinais e curativas: A saúde através das plantas. Vergara Brasil, Cotia.

Breffe, A.C.F., 2005. O Museu Paulista: Affonso Taunay e a memória nacional. Fundação Editora da UNESP, São Paulo.

Cabral, D. de C., 2008. Substantivismo econômico e história florestal da América portuguesa. Varia Historia 24 (39), 113–133.

Cabral, L.M., 2016. Plantas e civilizações: Fascinantes histórias da etnobotânica. Edições de Janeiro, Rio de Janeiro.

Carneiro, H.S., 2005. A farmácia tupiniquim. Nossa História 2 (21), 22.

Carvalho, M.M., 2002. A medicina em história: a medicina hipocrática. Saúde Mental 4 (1), 41–44.

Casagrande, D.G., 2000. Human taste and cognition in Tzeltal Maya medicinal plant use. J. Ecol. Anthropol. 4, 57–69.

Cordus, V., 1561. Historia plantarum. Gesner, Germany.
Corrêa, M.P., 1984. Dicionário das plantas úteis do Brasil e das exóticas cultivadas (6 v.). IBDF, Rio de Janeiro.
Cotton, C.M., 1996. Ethnobotany: Principles and Applications. John Wiley & Sons, Chichester.
Crespo, F.O., 1995. Fragoso, Monardes and pré-Chinchonian knowledge of Chinchona. Archeol. Nat. Hist. 22, 169–181.
Crespo, F.O., 1997. Monardes y Fragoso: dos protobotánicos del siglo XVI que se ocuparon de las plantas del Nuevo Mundo y las implicaciones de sus escritos sobre la introducción europea de la corteza del árbol de "quina" (Cinchona). In: Rios, M., Pedersen, H.B. (Eds.), Uso y manejo de recursos vegetales. Memórias del Segundo Simposio Ecuatoriano de Etnobotánica y Botánica Económica. Ediciones Abya-Yala, Quito.
Cunha, M.W., 2004. América de Américo: O desencontro de dois mundos em Cabo Frio (1503 – c. 1512), incluindo o naufrágio da Capitania de Gonçalo Coelho em Fernando de Noronha. Ágama. Armação dos Búzios.
Davis, E.W., 1995. Ethnobotany: an old practice, a new discipline. In: Shultes, R.E., von Reis, S. (Eds.), Ethnobotany: Evolution of a Discipline. Dioscorides Press, Portland.
de Caminha, P.V., 1500. Carta de Pêro Vaz de Caminha. Torre do Tombo, Gavetas, Gav. 8, mç. 2, n.° 8, Lisboa.
de la Calancha, A., 1638-1639. Corónica moralizada del Orden de San Agustín en el Perú. Imprenta Pedro Lacavallería, Barcelona.
de las Casas, B., 1967. Apologética historia sumaria. Edición preparada por Edmundo O'Gorman, con un estudio preliminar apéndices y un índice de materias. Instituto de Investigaciones Históricas; Universidad Nacional Autónoma de México, México.
de Mello, J.A.G., 1982. Manuel Arruda da Câmara – Obras reunidas. Fundação de Cultura Cidade do Recife, Recife.
de Sahagún, B., 1989. Historia General de las Cosas de la Nueva España. Editado por J. García Quintana e A. López Austin. CNCA-Alianza Editorial Mexicana, México, D.F.
Dioscorides, Pedanius, 1554. De materia medica libri sex. Apud Gulielmum Rouillium, Lugduni.
Dodebei, V., 2005. Memória, circunstância e movimento. In: Dodebei, V. (Ed.), O que é Memória Social? PPG em Memória Social da UFRJ, pp. 43–54. Rio de Janeiro.
Duniau, M.C.M., 2003. Plantas medicinais: Da magia à ciência no Brasil. Editora FIOCRUZ, Rio de Janeiro.
Fabrega, H., 1997. Evolution of Sickness and Healing. University of California Press, Berkley.
Fernandes, T.M., 2004. Plantas Medicinais: memória da ciência no Brasil. Editora FIOCRUZ, Rio de Janeiro.
Ferreira, A.R., 1971. Viagem Filosófica. Iconografia v.1: geografia e antropologia; v.2: zoologia. Conselho Federal de Cultura, Rio de Janeiro.
Ferreira, A.R., 1972. Viagem Filosófica. Memórias (zoologia e botânica). Conselho Federal de Cultura, Rio de Janeiro.
Ferreira, A.R., 1974. Viagem Filosófica. Memórias (antropologia). Conselho Federal de Cultura, Rio de Janeiro.
Ferri, M.G., 1980. História da Botânica no Brasil. In: Ferri, G.M., Motoyama, S. (Eds.), História das ciências no Brasil. EPU/Edusp, São Paulo, pp. 33–38.
Font Quer, P., 1967. Plantas Medicinales. El Dioscórides Renovado. Labor S.A., Barcelona.
Foucault, M., 1999. A Ordem do Discurso, fifth ed. Loyola, São Paulo.
Fragoso, J., 1572. Discursos de las cosas aromáticas, árboles y frutales, y de otras muchas medicinas simples que se traen de la India Oriental, y siruen al uso de medicina. Francisco Sánchez, Madrid.

Fuchs, L., 1594. Institutionum medicinae libri quinque. Per Hieronymum Gemusaeum, Basileae.
Galeno, 1550. Galeni omnia qvae extant opera. Apud Iuntas, Venetiis.
Garro, L., 1986. Intracultural variation in folk medicinal knowledge: a comparison between groups. Am. Anthropol. 88, 351–370.
Gerard, J., 1633. The Herball or Generall Historie of Plantes. Adam Islip, Joice Norton and Richard Whitakers, London.
Giannini, S.D., 2004. Santos Médicos, Médicos Santos. Panda, São Paulo.
Gottlieb, O.R., Borin, M.R.M.B., Bosisi, B.M., 1995. Chemosystematic clues for the choice of medicinal and food plants in Amazonia. Biotropica 27, 401–406.
Guerra, F., 1952. Códice Badiano: El manuscrito pictórico mexicano-latino de Martín de la Cruz y Juan Badiano de 1552. Editorial Vargas Rea y El Diario Español, Spain.
Herodotus, 2014. The Histories. Viking, New York.
Hinton, D.E., Howes, D., Kirmayer, L.J., 2008. The medical anthropology of sensations. Transcult. Psychiatr. 45, 139–141.
Hippocrates, 1555. Hippocratis coi medicorum omnium longe principis, opera quae apud nos extant omnia. Apud Ant. Vincentium, Lugduni.
International Society of Ethnobiology – ISE, 2006. Código de Ética da Sociedade Internacional de Etnobiologia. Available from: http://www.ethnobiology.net/ethics.php.
Janeira, A.L., 2006. Primórdios do colecionismo moderno em espaços de produção do saber e do gosto. Memorandum 10, 65–70.
Jeudy, H.-P., 2005. Espelho das cidades. Casa da Palavra, Rio de Janeiro.
Johns, T., 1990. With Bitter Herbs They Shall Eat it: Chemical Ecology and the Origins of Human Diet and Medicine. University of Arizona Press, Tucson.
Klanovicz, J., 2002. Uma etnobotânica no México. Episteme 15, 133–136.
Kleinman, A., 1978. Concepts and a model for the comparison of medical systems as cultural systems. Soc. Sci. Med. 12, 85–93.
Laws, B., 2013. 50 plantas que mudaram o rumo da história. Sextante, Rio de Janerio.
Leite, S., 1953. Artes e Ofícios dos Jesuítas no Brasil (1549-1760). Edições Brotéria / Livros de Portugal, Lisboa.
Lévi-Strauss, C., 1962. A ciência do concreto. In: Lévi-Strauss, C. (Ed.), O pensamento selvagem. Cia. Editora Nacional, São Paulo, pp. 19–55.
Lorenzi, H., Matos, F.J.A., 2008. Plantas medicinais no Brasil. Instituto Plantarum, Nova Odessa.
Luccioni, P., 1999. Um médecin botaniste, Dioscoride. La Garance Voyageuse 46, 5–8.
Marques, V.R.B., 1999. Natureza em boiões: medicinas e boticários no Brasil setecentista. Editora Unicamp, Campinas.
Medeiros, M.F.T., 2009. Inter-relações entre pessoas e plantas através dos documentos. Etnobotânica histórica: princípios e procedimentos. Nupeea, Recife, pp. 19–36.
Meneses, U.T.B., 1992. A História, cativa da memória? Para um mapeamento da memória no campo das Ciências Sociais, vol. 34. Revista do Instituto de Est. Bras., pp. 9–24
Monardes, N., 1571. Segunda parte del libro des las cosas Que se Traen de nuestras Indias Occidentales, Qué sirven al uso de la medicina; faire se trata del tabaco, y de la sassafras, y del carlo Sancto, y otras de muchas yervas y plantas, simientes, y licores Que Nueva Mente agora han venido de aqulellas partes, de virtudes y maravillosos grandes effectos. Alonso Escrivano, Sevilla.
Mosteiro de São Bento do Rio de Janeiro, 2004. Available from: WWW.osb.org.br/.
Murguia, E.I., 2010. Memória: Um lugar de diálogo para Arquivos, Bibliotecas e Museus. Compacta Gráfica e Editora, São Carlos.
Namer, G., 1987. Instituitions de memoire culturelle. In: Namer, G. (Ed.), Memoire et societé. Méridien, Paris, p. 242.

Nepomuceno, R., 2003. Viagem ao fabuloso mundo das especiarias, second ed. José Olympio, Rio de Janeiro.
Nogueira, E., 2000. Uma história brasileira da Botânica. Paralelo 15, Marco Zero. Brasília, São Paulo.
Occhioni, P., Arcoverde Sobrienho, M., Arcoverde, C.L.L., 1979. As plantas na medicina folclórica. Leandra 8–9, 199–210.
Orta, G., 1872. Colóquio dos simples e drogas e cousas da Índia e assim de algumas frutas achadas nela (várias cultivadas hoje no Brasil). Imprensa Nacional, Rio de Janeiro.
Palmer, C., 2004. *Plantago* spp. and *Bidens* spp.: a case study of change in Hawaiian herbal medicine. J. Ethnopharmacol. 24, 13–31.
Paula, S.G., 1998. Um inventário pioneiro de biografias para os historiadores das Ciências. História, Ciências, Saúde-Manguinhos 5 (1), 127–144.
Pelt, J.-M., 2003. Especiarias e ervas aromáticas: história, botânica e culinária. Jorge Zahar Editor, Rio de Janeiro.
Pereira, N.A., Jaccoud, R.J.S., Mors, W.B., 1996. Triaga Brasilica: renewed interest in a seventeenth-century panacea. Toxicon 34 (5), 511–516.
Pernoud, R., 1996. Hildegard de Bingen: A consciência inspirada do século XII. Rocco, Rio de Janeiro.
Peroni, A., Torry, B., 2007. Does the taste matter? Taste and medicinal perceptions associated with five selected herbal drugs among three ethnic groups in West Yorkshire, Northern England. J. Ethnobiol. Ethnomed. 3.
Phillips, O., Gentry, A.H., 1993. The useful plants of Tambopata, Peru: I. Statistical hypothesis tests with a new quantitative technique. Econ. Bot. 47, 15–32.
Pires, M.J.P., 1984. Aspectos históricos dos recursos genéticos de plantas medicinais. Rodriguesia 36 (59), 61–66.
Piso, W., Marggraf, G., 1648. Historia Naturalis Brasiliae: in qua non tantum plantae et animalia, sed et indigenarum morbi, ingenia et mores describuntur et iconibus supra quingentas illustrantur. Franciscum Hackium; Lud. Elzevirium, Amstelodami.
Pliny, 1855. The Natural History. Taylor and Francis, London.
Plotkin, M.J., 1991. Traditional knowledge of medicinal plants: the search for new jungle medicines. In: Akerele, O., Heywood, V., Synge, H. (Eds.), The Conservation of Medicinal Plants. Cambridge University Press, Cambridge, pp. 53–63.
Rodrigues, J.B., 1992. A botânica: Nomenclatura indígena e seringueiras. Fundação Andorinha Púrpura, Sociedade de Amigos do Jardim Botânico Rio de Janeiro, IBAMA. Instituto de Pesquisas Jardim Botânico Rio de Janeiro, Rio de Janeiro.
Santos Filho, L., 2005. Anchieta na história da medicina brasileira. Available from: http://www.novomilenio.inf.br/festas/anchie02.htm.
Schwartz, J., 2008. Herodotus Project. Available from: HTTP://www.losttrails.com/pages/Hproject.html.
Scliar, M., 2005. Uma guerra contra a morte. Nossa História 2 (21), 14–19.
Shepard, G.H., 2004. A sensory ecology of medicinal plant therapy in two Amazonian societies. Am. Anthropol. 106, 252–266.
Taube, K., 1993. The Legendary Past: Aztec and Maya Myths. British Museum Press, London.
van den Berg, M.E., 1993. Plantas medicinais na Amazônia: contribuição ao seu conhecimento sistemático, second ed. Museu Paraense Emílio Goeldi, Belém.
Van-Mensch, P., 1992. Towards a Methodology of Museology. PhD Thesis, University of Zagreb, Zagareb, Croatia.
Vellozo, J.M. da C., 1825. Florae fluminensis, seu, Descriptionum plantarum praefectura fluminensi sponte nascentium… Typographia Nacional, Rio de Janeiro.
Vellozo, J.M. da C., 1827. Florae Fluminensis… Icones Fundamentales. 11 volumes de estampas numeradas. Oficina Litográfica Senefelder, Paris.

Vellozo, J.M. da C., 1881. Florae Fluminensis. Descriptionum plantarum praefectura fluminensi sponte nascentium.... Archivos do Museu Nacional 5, 1–461.
von Martius, C.F.P., Eichler, A.W., Urban, I., 1840 and 1906. Flora Brasiliensis. s.ed., Monachii et Lipsiae.
von Martius, K.F.P., 1979. Natureza, doenças, medicina e remédios dos índios brasileiros, second ed. Imprensa Nacional, São Paulo. 1844.
Wagensberg, J., 2005. The "total" museum, a tool for social change. História, Ciências, Saúde-Manguinhos 12, 309–321.
Waldstein, A., 2006. Mexican migrant ethnopharmacology: pharmacopoeia, classification of medicines and explanations of efficacy. J. Ethnopharmacol. 108, 299–310.
Walter, J., 2000. História dos animais e árvores do Maranhão: Frei Cristóvão de Lisboa. Instituto de Investigação Científica Tropical, Lisboa.
Wüst, I., 1998. Continuities and discontinuities: archeology and ethnoarchaeology in the heart of Eastern Bororo territory, Mato Grosso, Brazil. Antiquity 72, 663–675.

CHAPTER FOUR

Conceptual model of historical ethnobiology

Contents

1. Considerations on the social memory of knowledge and actions

Social memory understood as a deposit of information about knowledge and actions of the past in relation to nature's elements means basing our interest in the forms of representation built socially and collectively. It is worth saying that the past, in this context is related not only to a distant past but also to a recent past including people alive with their expressions of memory through their orality or other kind of recent expressions. A fact or an event that took place a few minutes ago, immediatly becomes a memory of the past. Thus, memory is woven from successive and continuous representations that walk through time. The main objective is to study the representations of the forms expressed both through perceptible or imperceptible manifestations and how these representations refer to varied narratives seeking to understand the processes through which these transmissions communicate. Objects hold within themselves a body that has meaning in the ways of knowing, that is, in the production context and transmission of knowledge. In the relation between individual memory

Historical Ethnobiology
ISBN: 978-0-12-816245-3
https://doi.org/10.1016/B978-0-12-816245-3.00004-4

and collective social memory, objects and images reflect this interaction and they can be seen as the testimonies of this memory process.

At this height, it becomes pertinent to bring to light the fact that humanity is responsible for communication strategies when they bring positive and negative effects on the lives of its individuals, as well as for representing their ideas and own manifestations of their environment when they use different forms of expression. Among these several forms of expression or languages, there is verbal language, which is pointed as the first expression of humans and most utilized, when compared with all others. Taking into acount others forms of expressions, we find miming, photography, dance, music, painting, drawing, and writing, which constitute nonverbal languages (Coelho Neto, 2008).

When it comes to the object being concretized from written language, we can state that, like language and corporal gestures, writing is a cultural symbol constituted socially. Since each culture has its own symbolic system, symbols have some aspects intrinsic to their nature. Their unique characteristics are explained by the fact that they are understood by the people who know their abstract value, which makes symbols logical and coherent for that human group, and even so, they are still based on organization rules.

Notably, concerning written nonverbal language, one can find written records with available information about people's relations with the elements of nature that arose at the time when cultural complexity of human societies was, to some extent, in existence due to the emergence of more sophisticated technologies. From that moment onward, the human species began to express themselves through decorative, artistic items, and through writing as well.

Among the written records, there are the graphic inscriptions that include schematic drawings, prints, and rock paintings, and they are also called "simple images". The oldest graphic inscriptions date back to approximately 30,000 years BC, and they can be found in caves, rocks, and objects. They consist of schematic representations of various animals and objects. These representations are identified as rudimental forms of writing and, therefore, do not constitute a complete and structured graphic language system (Coelho Neto, 2008).

The proofs come from three regions: Mesopotamia (Sumerians, 3000 BCE), China (practically with no material change in a written rationalized system emerged in an independent structural way as early as 1000 years BC), and Mesoamerica (Mayas, 600 BCE). Over time, the writing system saw a reduction in the number of characters down to a few tens. At first, each symbol represented a concrete object or something that approached it (pictograms). Over time, ideas also came into the picture. A simplified signal was produced, in which the image is superimposed on the concept. In this way, ideas were related to the object (ideograms). In the wake of that came the phonetic indicators of values in that signals were used to designate syllables (syllabary, phonograms).

Along the history, in order to keep all the knowledge produced by humanity, the practice of registering this memory has been developed so that it could be preserved. However, it is worth mentioning that in this practice there was a hidden intention to monopolize individual and collective behavior, thus limiting social coexistence according to politics.

Through this practice, the process of listening, making the registers and organizing facts chronologically has developed as a way to document past situations that could be necessary to be remembered. Thus, documentary records were associated with the power of some and this exercise was maintained between different human societies. In this way, these records, functioning as documentary credibility of social memory, could preserve traces considered important to be evoked.

Speaking a little about the term "document" (term of Latin origin, *documentum*) it represents an object that has a propriety of conserving knowledge and facts. It also has other functions which are: education source, consultation source or proof document. Since, throughout history, documents have gained a central role in the structure of societies, and history pointed out that documents were also important in addition to oral testimony. During this process of documentary production another situation has been considered, that is, the need to think about spaces where these sources of knowledge created in different midias can form a collection and be preserved.

Then, "archives" (term of Greek origin, *arkheion*) came out with the proposal to be the guardians of documentary institutions which were born from writing. This act of writing was developed from the ability to exercise the thought as a technique required by the formation of the State (Murguia, 2010). In this context, the archive appeared as a space to store produced documents that were interesting for the exercise of power. In the beginning, archives and libraries functioned in a single space (Derrida, 2001). So, the archives could be thought as a space of memory where we could find instruments of the most varied types of narratives hearkened back to events, people, environments etc. Nowadays, archives can be understood by their memory functions related to history, literature, administration, among other aspects.

Considering memory, we can perceive that until the 17th century memory was valued as an art to remember and in that same period history emerges, as an expression of a counter-discourse in which the validity and verification were the main ideas to deal with documentation. A historical mark occurred some time after the period of the Second World War, initiating a movement focused on technology, mainly related to media diffusion. Another important situation occured at that time was a revisionist social attitude that provoked the development of projects related to the claim of minorities, including ethnic, religious, among others. What happened is that memory came to be considered of interest to provide the conditions for legitimizing the existence of political arguments, representations and cultural traces. In this way, a new dimension of history occurred, as it started to by focused not only on veracity, but also on meannings, feellings, representations, yearnings, etc. brought by individuals about the facts that affect their existence (Murguia, 2010).

In the quest for know-how from the past, which goes beyond time, we realize that in order to identify memory, whether individual or collective, does not give us the assurance that we fully hold it. In fact, we have memory locations or institutions where a variety of representations constituted by different types of bibliography, manuscript or printed documents, iconographies and objects can be found. Thus, we are confronted with testimonies or subjects/objects ("things") that we are looking for. These memory caches hold the potential to refer us to something, some information from the past (social actors, events, knowledge, and actions). And they start to make sense when we discover their liaison with an intellectual process with a set of constructed data. At an early stage, they are the things/testimonies themselves that will enlighten our trajectory to reach them. Following, they will give us the clues through an intertwining network that will, in summary, reveal possibilities for the interpretation of the human intellect. Hence, between things/testimonies and social memory there are numerous possible distances/paths to be covered without one having the correct paths to reach this union (Fig. 4.1). This is because social memory is not easily available to any conceptualization (Medeiros, 2016), but through clues many possibilities arise from the things/testimonies, and then we can develop conceptualizations about subjects, elements of nature or any other object of study.

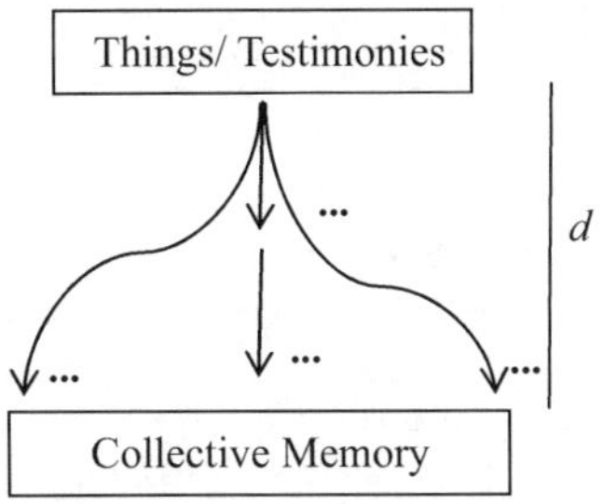

Figure 4.1 Graphic representation of the distances to be covered between things/testimonies and social memory for the constitution of discourse in historical ethnobiology.

In view of these considerations on the potential of testimonies as a source of access to social memory, let us begin to draw some lines of understanding, which will put us on track toward the things/testimonies. All this happens so we can attain the construction of an ethnobiologic narrative to bring to light the know-how and the facts of the past, namely, social memory.

2. Information as a documental source for memory representation

The existence of a close relation between memory and information, especially when we look at the history on sciences, letters, and the arts, can certainly be perceived along the history, and since the 20th century, memory and information have been considered the keys to understanding the world. Information and memory have, at the same time, an immaterial component (plan of ideas) and a material component (plan of concret representations of ideas) through which occurs the valorization of local communities with their heritages and cultural goods.

In this chapter, we intend to draw reflections and discussions on how these concepts work, information and memory, in an attempt at raising questions to contribute to the construction of their interrelationships. Our aim is to seek to establish a conceptual basis to broaden the dialogue scope between information and memory in the research field concerning historical ethnobiology. No doubt the dialogue spectrum is very broad, which may require us to create a boundary to limit the scope of the debate.

As for information, it can be said that humanity memory can be expressed as by orality, gesture, or graphic expression (writing or any other forms). There is a constant intercrossing of the individual memory with auxiliary memories represented by the collections safeguarded in spaces of

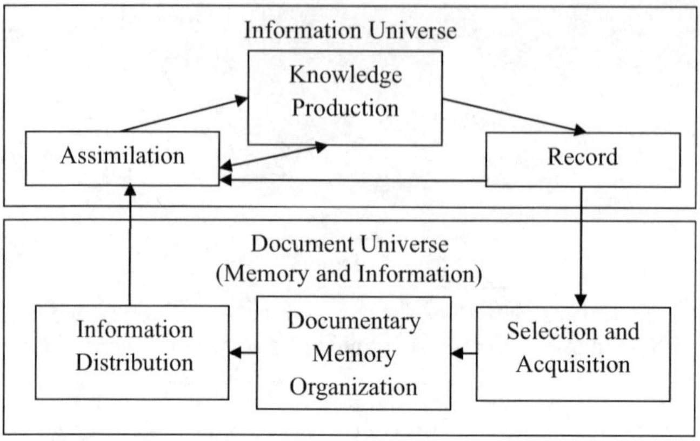

Figure 4.2 Graphic representation of the "information cycle". (That was *modified from Shannon and Weaver (1949)*).

memory (e.g. archives and museums). Nowadays, we find ourselves at a time when there is a great issue when we have memory and information living together, making changes, interactions each of other.

The set encompassed by the means that generate, safeguard and disseminate information/memories is, in its nature, represented by the amount of information and its representations through images. The process functions as a dispositive to memory and it has its own movement. A systemic model of the information life cycle was created by Shannon and Weaver (1949), the so-called "information cycle", which comprises two axes encompassing six moments, which are: knowledge production, record, selection and acquisition, documentary memory organization, information distribution and assimilation (modified from Shannon and Weaver, 1949, Fig. 4.2).

Let us briefly go through a few remarks on each one of the steps that composes the "information cycle."

2.1 Information universe

This axis of the "information cycle" describes the dimension in which information exchanges take place. The information universe is characterized for being a reality represented by the moments of production, record, and assimilation. These moments that compose this information universe are independent of the axis of the document universe. At first, there is the assimilation of knowledge (assimilation moment) where we can analyse the new knowledge produced (knowledge production moment) that is materially recorded (record moment) and that can be transmitted formally or informally (Fig. 4.2).

Let us now turn to the specific remarks on each stage of this subset, so we can consolidate the conceptual basis for the "information cycle" in ethnobiologic research with a historical feature.

2.1.1 Knowledge production

In this moment of "information cycle", we need to consider an aspect of making science nowadays. What we are experiencing today is that we have an interdisciplinary approach. Thus, due to the systematization of the knowledge, what is produced is no longer established in categorical areas of knowledge. Knowledge production can permeate multiple areas. This occurs not only in science, but also in the structure of the *corpus* of society. The boundaries would be defined by the action of the actor social who could find a range of meanings when searching for traces of memories (knowledge production).

This situation can be materially perceived from the experience that we had when we were researching the Monastery of São Bento Archives in Rio de Janeiro. We came upon manuscripts with monastery expenses during the 19th century. In principle, these documents would not be in the realm of our research, which was medicinal plants used in the Benedictine monastery of Rio de Janeiro at the time. However, while carrying out the search for information, it was found that the expenses were related to the material purchased for the apothecary (pharmacy), where the medicinal plants were prepared and the medicines were handled.

From this example situation, we concluded that documents with different social production origins could be connected from an act of reflection on the part of any researcher, who happens to be the one giving or failing to give meaning to the information that could otherwise be of interest only to those who were researching the 19th century economy. In this view, we can not have fixed fields of interest; rather, there should be a modular network that could be formed from the contact with documents produced for multiple functions.

2.1.2 Record

In the field of records we enumerate some supports where the knowledge (information) produced is available through a textual, sound, imagery, or three-dimensional register. These supports have been adapting and following the technology, which means that the information supports are diverse at the same time that they are in consonance with technology. They could also be categorized by a combination of these categories, as shown in Chapter 3. To the point, information supports have been evolving in the same pace as technologic advances. This aspect confers to the supports in information recording not only their diversity, but also their duplicity.

For this reason, nowadays, for instance, knowledge can be accessed in its original and classic vehicle, both on paper and in current media (magnetic or electronic). We can also use as an example the access to an event about a product such as the scientific exhibitions titled "Monks' Botanic Pharmacy", held at the Monastery of São Bento de Olinda, in 2011 in the city of Olinda (Northeastern Brazil), and also at São Bento Monastery in Rio de Janeiro, in 2013, in Rio de Janeiro (southeast region of Brazil). These exhibitions showed a diversity of supports in their conception by interweaving original manuscripts, photographs, cartography, plant specimens, and thematic videos on the history of conception and the activities of the *botica* (pharmacy) by the Benedictine monks of these two cities.

Although we have remarked before about the great variety of supports, they do not always occupy the first place in the research process when the focus is in the theme. But, there are situations in which the support becomes the object of investigation, and the theme is no longer the main object of study. Thus, support becomes the main objective (purpose) of the research, and is not just an accessory to find sources. Therefore, there may be an inversion of the focus of importance between the support and the theme.

2.1.3 Assimilation

This moment comprises the assimilation of information generated by knowledge, which will provide the production of new knowledge.

In this moment, considering the sense of postmodern humanity, we can realize that knowledge has a dynamic inserted in the real time. The permanence and change of knowledge are adequate and follow step by step the speed of the transmition of information that arrives to us in real time (Virilio, 1994).

The phenomenon we are experiencing is characterized by the speed information (or products) presented to society. In this way, behavior of society is established in relation to knowledge that is modeled by the velocity of information causing a constant expectation of people to acquire such knowledge. And some questions arise as to whether this information would be updated, the most recent, or already out of date? The person's feeling can be identified as an urgent necessity to be well informed. It is an expectation provoked by the realization that there is important new information, but not yet accessible. And also, there is another expectation generated by having produced some relevant information and not making it visible.

Oftentimes, this type of expectation occurs when one is working on a theme, and before long, one wants to find the answers to questions, such as where will we find the documents pertinent to this subject? Who could point us to the right direction? How are we supposed to work the information accessed (types of transcription, spreadsheet forms, and analysis)? Still, there is also that type of expectation resulting from knowing the location of information relevant to the topic and also the focus of the research, but the access to that documentation for a variety of reasons is hindered, for example, by not being able to access private collections, or when that is not the case, the document is in an advanced state of deterioration, there being loss of information as a result.

To precisely state which information is to be collected and how much of that information must be assimilated is an issue that has to do with complexity. And it can be measured only indirectly by the quantity and quality of the new knowledge produced, and when one is in the field proper, in close proximity with the information, the memory, and the collection.

2.2 Document universe (memory and information)

The document universe (memory and information) comprises the second axis, tied to the first axis called the information universe, on which it depends. Neverthless, as we can see in the Fig. 4.2, the memory in itself only can be conceived through selection and acquisition, documentary memory organization, and information distribution, which are the three moments that will be characterized below.

2.2.1 Selection and acquisition

Documental memory can be investigated in two ways. A search mode is the traditional uniterm tabs. The other way is to explore documental memory electronically utilizing the structure of search models of bibliographic databases. So, these two paths are the constitutive conditions that we have to locate the documentary sources that will help us in historical biological researches.

2.2.2 Documentary memory organization

There are a great deal of definitions for the representation of knowledge. Among several, there is one that considers the representation (documentation memory organization) as a construction from the perspective of the social actor (Gomez, 1993). We will also find another one in which the meaning of the representations is based on the concept of the interface of mediation between social actors and their environment. There is another, in this reflexive construction of the meaning of representation, which is to consider the individual and collective information transmition in which there is always a mediator between those who communicate and those who receive. This mediation comes to life through the accumulation of information. By adopting this sense, we can apply the expression "representation of information" created by Kobashi (1996).

2.2.3 Information distribution

The information distribution comprises the dissemination of the knowledge. This process integrates the structure of this document universe and also constitutes a natural characteristic of the human being, since we use the dissemination of information to communicate our stock of information. This distribution, dissemination or transmission of information can be understood as the interaction between the acquisition and the representation of documentary memory.

We can pave the way to establish the relations between the representations, by establishing a theme generator, a key type or from a source/document, which will be relevant to our research focus. It means to say that what is represented in the documental memories is gradually being unveiled by the researchers, who will determine the information items pertinent to their search.

Thus, we can perceive the formation of a network through the various possibilities that can be proposed on the various representations of documentary memory. It is a process that stimulates the information distribution of a specific item, going beyond its material form and its content, giving social actors a *spectrum* of items that can be obtained according to their needs of information. Besides that, information dissemination also has strategies for limiting information so that unwanted ones do not appear using filters. This is the key to the organizaton and use of guides and catalogs structured by themes, areas or even domains of knowledge. These search engines are essential in the representation-seeking process, and they differ from searching lists by terms (traditional search lists).

3. Proposition of a conceptual model for historical ethnobiology

Next we will develop the conceptual idea of making knowledge through historical ethnobiology as a conceptual model.

The conceptual model that we have developed is based on the dynamic transfer of information from the document, which comprises past preserved evidence for the formation of the ethnobiologic scientific narrative (Medeiros, 2016).

The model followed was the representation of the information cycle derived from the interrelation between the subject and the object, taking into account the spatial, environmental, and temporal (including time) levels (Medeiros, 2016). These elements were considered essential constituents of a social memory that may be useful for the construction of an ethnobiologic narrative endowed with an internal coherence.

Below, we have structured a theoretical-conceptual model to research on historical ethnobiology, so we can construct shades of intersection between memory and information. In this sense, the process of documentary analysis for historical ethnobiology takes as its basis the relation between its constituent elements represented in a schematic form soon to follow (Fig. 4.3).

This proposition is based on the assumption that the focus of study that deals with object—culture—memory is interdisciplinary in nature. Hence, the methodologies or technical procedures of investigation, dissemination,

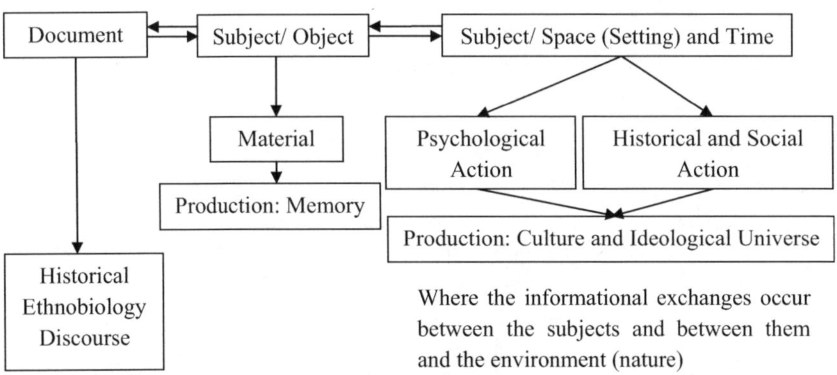

Figure 4.3 This scheme organizes the memory cultural field in its approach to information, in the constitution of an ethnobiologic discourse. (*This scheme was expanded through the model proposed by Medeiros (2016)*).

and protection of information, which are adopted by a given discipline, are not sufficient to conduct interdisciplinary research. The contradictions inherent to each discipline will emerge as a sign of theoretical and methodological insufficiency for the interdisciplinary field, or as a counterpoint between the areas involved in a particular research.

The unique views of each discipline from which historical ethnobiology utilizes will be part and parcel of this research with focus on interdisciplinary approaches. Therefore, this science shows a special feature on seeking an articulation of human sciences with the sciences of nature, that is, the study of nonsingular questions. Upon analyzing the characteristics of questions, Morin (1977) analyses that the nonsingular ones follow the principle of disorder, obscurity and complexity due to their wide scope and multiple manifestation. Thus, what is intended in the study of nonsingular questions on historical ethnobiology is that from the principle of complexity: possible connections between memory fragments are proposed through the construction of ethnobiologic narratives.

The idea developed here was to present a conceptual model of the phases involved in the research on historical ethnobiology and, from the comprehension thus gained, prepare the reader for the next chapter (Chapter 5), where we will be discussing the methodological procedures employed in this research without emphasizing oral memory as we have seen since the previous chapters.

References

Coelho Neto, A., 2008. Além da revisão: critérios para revisão textual. Editora Senac, Brasília.

Derrida, J., 2001. Mal de arquivo — Uma impressão freudiana. Relume Dumara, Rio de Janeiro.

Gomez, M.N.G., 1993. A representação do conhecimento e o conhecimento da representação: algumas questões epistemológicas. Ciência da Informação 22 (3), 217—222.

Kobashi, N.Y., 1996. Análise documentária e representação da informação. Informare 2 (2), 5—27.

Medeiros, M.F.T., 2016. Historical Ethnobiology. Introduction to Ethnobiology. Springer, Switzerland, pp. 19—24.

Morin, E., 1977. La method: la nature de la nature. Édicions du Seuil, Paris.

Murguia, E.I., 2010. A memória e sua relação com arquivos, bibliotecas e museus. In: Murgia, E.I., Org (Eds.), Memória: Um lugar de diálogo para Arquivos, Bibliotecas e Museus. Compacta Gráfica e Editora, São Carlos, pp. 11—32.

Shannon, C.E., Weaver, W., 1949. The Mathematical Theory of Communication. University of Illinois Press, Urbana.

Virilio, P., 1994. A máquina de visão. José Olympio, Rio de Janeiro.

CHAPTER FIVE

Methodological aspects for researching in historical ethnobiology

Contents

1. Making ethnobiology science through historical documents

In this chapter dedicated to the methodological aspects of research in historical ethnobiogly, we will make considerations that will help us to structure the description of field situations that involve indirect documentation that we normally encounter. We will bring guidlines in the first and second topics, where we find ideas that we had the possibility to develop in previous situations and will be expanded here (Medeiros, 2009, 2016). Concepts that we believe to be of fundamental structural order when considering the core of this methodology. Thus, let us start this first topic by thinking: How can we make science by accessing historical sources?

We believe that so far enough evidence has been shown to prove how worthy the documents produced throughout the history of mankind are. Equally, it is already clearly seen how these historical sources are paramount

Historical Ethnobiology
ISBN: 978-0-12-816245-3
https://doi.org/10.1016/B978-0-12-816245-3.00005-6

for the study and understanding of human society's constitution and their interaction with nature.

Then, this question will now be developed from the point of view that making science through historical sources requires the ethnobiologist to have interactions with documental sources. This is the basis for the development of research with a historical perspective. For it is precisely within the preserved vestiges that the desired raw material can be found, which is indispensable for the reconstitution of the past, and the construction of structured projections for the future. In this sense, it is important to emphasize that the construction of an ethnobiologic narrative is part of the historiographic operation developed by the ethnobiologic researcher.

The starting point of a research on historical ethnobiology is not in the analysis of a document; rather, it is in how questionings are formulated, and in how documentary sources are problematized. By problematizing the sources, structuring steps are paved in the historiographic operation. Problematizing becomes a central action in the research since the sources represent the testimonies who will answer the issues formulated and presented to the sources by the ethnobiologists. Pursuant to this line of thought, Lopes and Galvão (2001) will say that the definition of what sources are like is determined by the problem that the researcher problematizes. Le Goff (1994) refers to problems in historiographic research as the movement toward how documents are questioned.

Hence, the questionings put by the researcher in regard to the document are as important as the document itself. This is to say that the problems raised by the ethnobiologist about the document are what also lend meaning to it.

Therefore, this movement of the ethnobiologists' investigative act leads us to question the sources as being the accumulated narratives where many historical aspects on varied themes are constructed. In this feature, we can postulate that the document becomes a monument. The document assumes this representation because it is the result of the effort of historical society to choose and preserve for the next generations documents that will reveal in the future a specific image that they would like to express.

This current action of dealing with documentary sources transforming documents into monuments (Le Goff, 1994), and then presenting them as elements to be seen as a single piece, and then as a new form of grouping, gives new meanings when compared to each other.

The utilization of sources has been expanded, which created space for the incorporation of information present in any trace or vestige left by past societies (see Box 5.1 for the definition of source). It is then possible

Box 5.1 Definition of documental sources.

Documental source definition

Source is all evidence of the past (written, iconographic, oral source) "that men and time have voluntarily or unintentionally preserved, and that the [...] [researcher] consciously, deliberately and justifiably decides to erect elements to reconstitute a particular sequence of the past, to analyze it, to restore it to its contemporaries in the form of a narrative, that is, a writing endowed with an internal and irrefutable coherence, therefore of a scientific intelligibility".

Adapted from Rousso (1996, p. 86).

to use documental sources as diverse as possible for writing a narrative in historical ethnobiology.

The sources and types of documents in documental research may include primary sources, which comprise documental research, and secondary sources, which encompass bibliographic research. The documentary is restricted to documents. On the other hand, bibliographic references cover the secondary data (bibliography) about a subject of study already made public and put the researcher directly in contact with all the existing material on the subject. Data sources cover both written and nonwritten documents, as well as those made at the time of occurrence or later on (contemporary sources or retrospective sources). Then, a survey of all sources that may be useful is in order, namely dictionaries, which define the meaning of words; encyclopedias, which provide an overview of the subject; books, which develop the theme; magazines/newspapers in search for more specific articles; the internet, to collect images and information in general; and interviews/surveys for the collection of opinions and statistics.

Therefore, there are multiple possibilities for documental types that may be useful for research on historical ethnobiology. Documental types are commonly associated with books and printed materials. Although, there are some interesting derivations in this very typology that can be briefly commented upon. It is not uncommon to come across handwritten samples whose handwriting is extremely difficult to read and understand. A very striking example of such a sample is the writings left by the Brazilian journalist Assis Chateaubriand, who is said to have recognized that his handwriting was illegible. In a letter addressed to Plínio Barreto, Brazilian editor of the magazine *Revista do Brasil,* Assis Chateaubriand himself recognized the effort to understand the reading of the manuscript of his authorship sent to this magazine (Chateaubriand, 1917). It is also noteworthy, the technologic change involved in the production of documental vestiges by mankind. For instance, still commenting about the journalist Assis Chateaubriand, this advance contributed to having access to his letters, and other types of documents by people, since they started to be typed and no longer handwritten.

However, documental types will also cover watercolors and drawings. An example of this documental type can be found in Jean-Baptiste Debret's 19th century drawing (Fig. 5.1). This drawing is a preparation for engraving on white paper, with Indian ink, depicting indigenous people of different ethnic origins in Brazil.

Documents liable to be investigated also include art objects such as the illustrated edges of book pages through the refined painting technique of the 18th century. This decoration in the books had different motifs, for example, some angles of the city of Rio de Janeiro, Brazil, which were hand designed on the edges of the pages (see *The Modern Traveler.* Fore-edge watercolors performed on the frontal cuts of each of the six volumes;

Figure 5.1 Botocudos, Buris, Patachos and Macharis (1834). Jean-Baptiste Debret, from *Voyage Pittoresque dans le Brésil*, sketches, 32.2x47 cm (open), belongs to Pinacoteca do Estado de São Paulo collection, São Paulo, Brazil.

London; Trudy Lowndes; 1777; 6 volumes) (Anon., 1777). This technique is called fore-edge painting, whose visualization of the illustrations is becomes visible only by tipping the leaves.

In this array of documents, we have also found printed iconographic albums, lithographs, and photographs. Lithographs mark the reproduction of photographs early in the history of photography, and that visual resource was used to minimize the appearance of photography, which at the end of the 19th century will be employed in the production of illustrations.

As it has been explained in depth so far, the beginning of research on historical ethnobiology is then the establishment of an investigative problem, which was thought to be the guiding principle for the collection and selection of documental sources. Problematization will lead the researcher to initiate his or her investigation by choosing an element of nature or a subject to be approached. Following this initial procedure, one starts to identify the sources that will be indispensable to the development of the project.

Most importantly, however, it should be noted that there is still the possibility of starting research by taking the historical documentation as the starting point. It is probable that the researcher will come upon some documentation through the work dynamics, which will provide him with the glimpse of a new investigation.

Therefore, through the establishment of a theme generator, key species, or key source, we will be establishing relationships between the representations relevant to our research focus. This is to say that what is represented in the documental memories is gradually being unveiled by the researcher, who will ultimately determine the information items that will be applicable in the research.

In historical ethnobiology, documental analysis becomes the method to be adopted in its research. This methodological procedure credits the document with a centrality in the research process; that is, it is in it that the information necessary for the researcher's analyses is to be found. This documental research encompasses two stages, namely, the collection and analysis of documents.

Document collection includes some substeps. The first is to assert the memory locations where the materials to be identified and consulted are safely kept. Memory locations is presented under two optics: that of the symbolic space represented here, for instance, by the name of streets; and the physical space, which includes libraries, archives, museums, monuments, scientific collections, and others. These memory locations also have two related aspects: the symbolic and informational ones. The symbolic aspect stems from its representation within a space through which human society is able to remember its social memory. Whereas, the informational aspect happens through evidence when the recall of events, people, and places are allowed.

We move from the conception planning of ideas, depth, adjustments, sizing, location, and identification to selection and acquisition. Selection can be achieved in a traditional way ("uniterm" tabs) or electronically (inverted serial files); the latter is present in search models for bibliographic data. As we begin the selection of sources, it should be crystal clear that a selection has already been made either by time itself, by those who produced the documental material, by those who have preserved it, by those who eventually left some sort of intentional ruin in the material, and still, by those who organized the collection.

Documental acquisition is dependent on a principle highly required of a researcher, that is, a thorough and detailed assessment of the records to be consulted. This principle aims to analyze the general access conditions to the documental sources. It is at this stage that ethnobiologists should make sure there is consultation availability for the research in the memory location where he/she found the sources intended for utilization. Since access to documents is not always conceded, to know about this availability is of paramount importance (Quivy and Campenhoudt, 2003). Therefore, to well develop the research, it is desirable to develop good mediation with those responsible for the institution's collection (memory location) by publicizing the research project and saying about how the collection of this institution could help you. Thus, this is an essential practice to be adopted during this process.

In this sense, a word to the wise is in order regarding the necessity to verify the general conditions for admission of the consultant for both the selection and the acquisition (researcher) (Khoury, 2004). It means that we must educate ourselves in the aspects involving our fieldwork, during the data collection in the memory locations. Therefore, we must be concerned with whether we are allowed to reproduce the material in consultation, borrow it for reproduction, what forms of reproduction are allowed (whether electrostatic, photographic, videographic, digital, etc.), and which ones will be the financial responsibility of the institution, and which ones will be the researchers'.

Other aspects related to the selection and acquisition of information include knowing which are the existent research tools, such as catalogs, guides, binders, and listings that report on the collection of the institution under consultation (Table 5.1). These research instruments comprise reference works to identify, locate, summarize or transcribe archives, funds, groups, series, and documentally pieces for the purpose of systematization and access to the collection. These instruments should provide the elements

Table 5.1 Archival description of documentary collections, based on CONSELHO INTERNACIONAL DE ARQUIVOS (2001) CONSELHO INTERNACIONAL DE ARQUIVOS (2004) CONSELHO NACIONAL DE ARQUIVOS BRASIL (2006).

Instrument	Description
Summary archival description (Reference sets of documents such as background, group, subgroup, or series)	
Guide	- Displays information about the history, services, and funds[a] of one or more archives
Source guide or research guide	- Identifies and finds funds, groups, or series[b] of one or more archives, referring to a particular theme
Inventory	- Describes the series very thoroughly, expressing the links between the documents
Archival description analytical (Reference documentary pieces)	
Catalog	- Its description unit is the documental piece[c]
Selective catalog	- Describes preselected documents, which belong to one or more funds or archives
Editing texts, editing documents, or editing fonts	- Transcribe the document in its entirety
Index	- Systematic indexing product that allows the retrieval of information present in the documents

[a]Fund: principal and structural arrangement unit in the files consisting of documents originating from the same source generating the files, as well as more than one source generating files, including public or private entities, person or family, in the exercise of their functions and activities, united by the similarity of their activities, and respecting the provenance. This unit brings together documents that have organic relations between them and are preserved as proof or legal and/or cultural testimony, and should not be mixed with documents of a different documental set generated by another institution, even if it is related to it (Bellotto, 2006). It can also be called documental fund or archive fund.
[b]Series: documents organized according to the filing system and kept as a unit, because they result from the same process, accumulation or archiving, or from the same activity, have a particular form due to some other type of relationship derived from the production process, in receiving or using. It is also known as a series of documents.
[c]Documental pieces: document that represents the smallest documentary unit, which is intellectually indivisible, integral to dossiers or processes. It constitutes a documentary unit that is physically indivisible. It can also be called documentary item.

needed by the applicant to carry out a consultation with the collection keepers. Thus, they must give functional indicatives leading to the series of documents present in the collection (CONSELHO NACIONAL DE ARQUIVOS BRASIL., 2006). They enable efficient, agile, and comprehensive access to the collection through descriptive programs. For the establishment of descriptive programs, the custodial institution normally seeks to establish the profile and quantity of the collection under its custody, the physical state of the documents, the institution's human and financial resources, and in addition to these descriptive aspects, the users' profiles and their information demands are also queried.

It seems so natural and obvious, but no less important for the planning and optimization of the fieldwork to find out the days and time schedules the research unit operates. It is also worth mentioning that being informed on the conditions for the admission of the consultant will help us to better define the data collection procedures, so we can explore all the potential offered and allowed by the documental storage place. In this way, if we are to have photographic reproduction of the material, we must plan this activity a few days in advance, according to the material needed. We highlight this photographic imagery example so we can emphasize that images, in general, are documents to prove a fact or phenomenon; and they are the expression of a temporality, identity and memory.

In the specific case of photographic records, they are conceptualized as a memory medium that displays visual narratives, for example, a landscape view about rural or urban areas, special social moments of people, such as specific or general celebrations, and elements of nature revealed through images. One of the main features of photography is to allow the perception and visualization of the changes (Lima, 2007), that occur either in the environment, in the life of an individual, in a research institution, or in the most varied themes of interest of those who adopt photography as a means of language. Pursuant to this feature, the set of images will confer chronological ordering and access to the most varied angles of perception on the same topic that is being focused (Lima, 2007). As a result, the use of visual evidence is complex in its analysis and, at the same time, indispensable for the composition of data research, either in the preparation of documents dedicated to scientific communication or in the organization of the social feedback from the project.

Another important aspect for the researcher's reflection lies in thinking how to access information. On accessing a certain documental material, we ought to be utterly attentive to the issue of assessing the collection's state of conservation. From this assessment, the researcher must then consider whether the progress of the consultation with the sources should continue, as this will be done through a systematic treatment, from where the answers to their research questions will be collected. This willingness to assume responsibility concerning access and documental handling stems from the fact that, in some place of custody, the material will be in different conservation stages, such as an intermediary or advanced deterioration process. It is then clear to the researcher's perception that the material should undergo an indirect intervention, which corresponds to a cleaning process, or a direct one to include a deeper intervention of restoration of the information support. The maintenance and storage of collections in an ideal environment require constant investments to keep the place for the most part dark and permanently cold and dry; although, these conditions more often than not cannot be met. As a result, the researcher ends up finding a much-degraded material and he/she has to take a stance towards this reality at the time of the research.

Therefore, besides the need for handling and consultation of the document, the document conservation should be a first concern as this will be affected by the systematic handling. Hence, a safeguard is in order regarding the care for the handling of the sources because this is a mechanical action and said to be the systematic action that causes damages to the documentation. This is to say that inadequate handling may cause physical and chemical damages to the collection by impregnating the document with sweat and dirt (Cassares, 2007). To mitigate these mechanical and chemical actions during consultation, certain practices are essential. These include the consultant's approach by utilizing disposable surgical gloves, masks, and goggles as needs arise. We must once again emphasize another important point regarding a proper posture when consulting collections, which is not to perform the information transcription by placing the paper on the document if a manual transcription is being done. Another relevant consideration is the act of forcing the opening of pages of the documents, what is unacceptable. In these cases, for a better reading and adequate handling of the document, support materials such as book holders or cushion pads should be required to the document custodian.

With these considerations in mind, let us carry on and talk about the fact that besides the collection stage, document analysis also undergoes a few substeps. Critical analysis will lead us to a group of operations and verifications, which emerge from the reading of documents to assign those meanings according to the research importance set by the researcher.

The descriptive structures of the data collected must take into account the need to adapt a general methodology to particular situations, in which case, there may be variations in the composition of the data sheet and in the information analysis. To set the data collection mode will depend on the types of sources we selected to our research. Hence, the guiding principle should be to keep in mind the information being sought, so it can be followed up with more detailed and systematic readings of the document. It is still interesting to look for a certain information collection and transcription pattern for an optimization of the work time in the field and, later on, the work in the laboratory when organization and data analysis and the writing of the final research document will be done. This data transcription pattern should look for a group of elements in common (or categories) that will compose the database already in the laboratory stage. We can adopt a collection of worksheets and cards that will allow us to create a data collection pattern and, then, to analyse our results in an appropriate way.

Considering the process of reading and transcription, Werneck et al. (1993) presents two ways in which this procedure can be performed specifically on the element's transcription contained in the documents: line-by-line or running transcription. The line-by-line procedure views the information content of each line of the document, with the numbering process of each line by the researcher. There is no concern in respect to the content expressed by the author of the document in each of the lines within the running transcript. To make sure the proper way for transcription is being utilized, the researcher should follow the technical norms for the document's transcription, especially the manuscripts. As these standards seek to establish the criteria of paleographic editions, they enable a presentation that follows uniformity in terms of spelling, conventions, signatures, and public signs for the analysis of mixed documents, for seals and stamps, and also graphic presentation for transcription (see Table 5.2).

A term that must remain as written in the consulted original document is the popular name by which a certain element of nature is designated as a reference in the content. Thus, the textual element transcription that preserves the original spelling of names, for example, of plants and animals, is a thorough practice in the collection of information. To achieve the real meaning of important terms that describe elements of nature, dictionaries and other secondary sources of that period, or also a linguist can be consulted.

Still another step essential in a research is documentary analysis (Table 5.3). It is intended to account for the document biography, which

Table 5.2 Some criteria for standardization of paleographic editions, especially for manuscripts, based on Houaiss (1990).

Criterium	Description
Spelling	- Separate words spelled unduly together and unite the syllables or letters spelled separately, but unduly. The joints of the proclitic (madê, selhedê), mesoclític and enclitic pronouns are excluded from the verbal forms on which they depend (meteremselhe, procuraase) - Write the letters in the usual way, regardless of their phonetic value - The double caudate "s" should be transcribed as "ss" and the simple one as "s" - "R" and "S" with sound of "rr" and "ss" should be transcribed as upper "R" and "S" - Ramist letters "b," "v," "u," "I," and "j" should be kept as they are in the document - Roman numerals should be reproduced according to the shape of the time - To the mistakes, omissions, repetitions, and truncations, which compromise the understanding of the text, the Latin word sic must be used between brackets and highlighted: [sic] - Noncurrent abbreviations presently should be developed with tap accruals - Abbreviations that are still usual or easily recognizable may be retained - Special signs of Latin origin and monogrammatic symbols and words will be unfolded, for example, & r^{a} = etc.; IHR = Christus - Signal of nasalization or tilde, when with value of "m" or "n" should be maintained - When there is a doubtful paleographic reading of a term, an interrogation must be placed between brackets after it: [?] - The original accent should be maintained - The original punctuation should remain the same - Uppercase and lowercase letters should be retained - Spelling must be maintained in its entirety, with no grammatical correction
Conventions	They indicate accidents in the original manuscript, such as illegible or damaged writing - Words that are partially or totally illegible, but whose textual meaning allows them to be reconstituted, should be transcribed in brackets

(Continued)

Table 5.2 Some criteria for standardization of paleographic editions, especially for manuscripts, based on Houaiss (1990).—cont'd

Criterium	Description
	- Illegible words should be indicated with the word illegible between square brackets and italicized: [*illegible*] - Lines or words damaged by the corrosive action of paint, moisture, tearing, or corroded by insects or other animals should be indicated in the transcript. For example, it should be indicated by the expression corroded between square brackets and with the approximate mention of its extension number: [*corroded* $\pm$ *6 lines*] - Interlinear or marginal autographs that complete the writing should be inserted in the text between the signs < … > - When they are not autographs, they will be indicated in footnote - Marginal notes not inserted in the text should be kept in their place or in sequence to the main text with the indication: the right margin or the left margin - Other hand notes must be transcribed in footnotes
Signatures and public signs	- Signatures in shallow or rubrics will be transcribed in italic font - Public signs shall be indicated between square brackets and italic font: [*public signal*]
Mixed documents	- Printed characters that appear in mixed documents (for example, forms, stamps, standard sheets, stamps, acronyms) should be transcribed in different types
Stamps, stamps, etc.	- Stamps, signets, sealing wax, sealed papers, and designs shall be indicated by their nature in brackets and underlining: [*stamps*] - Printed sayings and the value of the stamps will be transcribed inside square brackets and underlined: [*stamps*]
Graphic presentation	Transcription may be line-by-line or running transcript - Respect the original document paragraph division - Number the pages according to the original document, always indicating the change of each one, between brackets and in the middle of the text, including the verse: [fl. 5], [fl. 5v] - If the original is not numbered, the transcriber must number it. The numbers added will be inserted in the transcription between brackets and in italic font: [*fl. 3*], [*fl. 3v*] - Blank sheets shall be indicated in square brackets and italic font: [*fl. 9, blank*]

Table 5.3 Biography of the textual and imagery document required during data collection and processing, based on Gastaminza (2005).

Relative questions	Related attributes
Biographic data: modal competence	
When?	Dating *Time coordinate*
Where?	Location *Space coordination*
Biographic data: technical competencies	
What are the material characteristics?	Technique and dimensions (height and width) *Material characteristics of textual and imaging document support*
Biographic data: production context	
Who did it?	Authorship
Biographic data: theme	
What for?	Subject
Biographic data: encyclopedic competence	
What is it?	Content *Set of information to identify the content*
Biographic data: iconographic competence	
What is it?	Forms *Ability to identify visually expressed forms*

includes information on modal competence, technical skills, production context, theme, encyclopedic competence, and iconographic competence for visual content (Gastaminza, 2005).

It is fundamental, for example, that the biographic information be collected, since they will compose the descriptions and captions of the imagery records. For the use of this type of document, we must have include the guardian's authorization regarding such reproduction recognized authorship, as well as the authorization for the use of third party images. The indication of the bibliographic reference is also required. For the organization of the bibliographic elements, we must adopt the standard determined by each institution or publication during the writing of the scientific document. It should also be noted that as for handwritten and printed sources, the elements that indicate the collection origin and location of documents need to be stated.

In general, we can list some sources, some documental types, and suggested analysis procedures for each, as shown in Tables 5.4 and 5.5. These

Table 5.4 Document sources, documental types, and procedures in documental research based on Lakatos and Marconi (1991).

Research type: documentary (primary sources)

Document sources/types of documents	Description
Public archives (National, state, municipal)	
Official documents	- More reliable data source - They express individual acts and acts of political life - There should be an interpretation and comparison of the consulted material, since it is not known how these documents were created - Documentary examples: royal orders, laws, official forms, reports, annuals, permits, etc.
Parliamentary publications	Records from the activities of government agencies (chambers and senate) - Reliability hardly questioned - Documentary examples: minutes, debates, bills, reports
Legal documents	- Source of sociologic information - They demonstrate the regulation of social behavior and how social problems are formed - Documentary examples: records of births, marriages, deaths, wills, etc.
Administrative publications	- They are less trustworthy than previous documents - Represent external organization (with the general public) - Represents the administrator's philosophy (when the document is for internal use) - A study of the political moment is necessary for a good understanding of the activities described

Table 5.4 Document sources, documental types, and procedures in documental research based on Lakatos and Marconi (1991).—cont'd

Research type: documentary (primary sources)	
Document sources/types of documents	**Description**
Private archives (Private households*, private institutions** (schools, churches, associations, etc.), public institutions (health clinics, police stations, etc.)***	
Private documents	- It contains informational content that describes the meaning the facts bore for those who experienced it - Documentary examples: *correspondence, memoirs, journals, autobiographies, etc.; **records, trades, correspondence, minutes, memorial, programs, comuniques, etc.; ***data on hospitalizations, diseases, etc.
Different private and official institutions (Federal, municipal, and state statistical departments, etc.)	
Statistical sources	- Collected directly - Collected periodically - They can encompass the entire population (census) or part of it (sampling techniques), expanding the results for the population as a whole - Encompass on population characteristics (e.g., age and schooling); factors influencing population size (such as births, deaths, illnesses, etc.); economic factors (such as active labor force, unemployment and per capita income); housing (such as number of dwellings, number of rooms, infrastructure such as light, water); and the media (such as radio, telephone and television)
Distributed in previous sources	
Iconography	- Only source of testimony on various aspects - They allow the reconstitution of a setting, the verification of trends, etc. - Documental examples: every image represented by engravings, prints, drawings, paintings, etc.
Photographies	- They have the same vocation as iconographies - They depict a recent past
Objects	- They reveal the level and meaning of evolution - They reveal the meaning of the object symbol in a particular social context

Table 5.5 Documental types and procedures in bibliographic research based on Lakatos and Marconi (1991).

Type of research: bibliographic research (secondary sources)

Types of documents	Procedures
Written press	- Data from newspapers and magazines should be analyzed for the following: Independence Check the theoretical assumptions held by the countries the material came from, whether totalitarian or democratic, and also the advertising agencies regulation modes Content and orientation To perceive the spaces dedicated to the subjects dealt with in the publication and how the issues raised are dealt with Diffusion and influence To identify the region as it is distributed and the targeted population Interest groups Publications targeted to specific groups, such as professional categories In these publications' ideas, interests and performance of the representatives in this social group can be perceived
Audiovisual media	- Content of audiovisual media, films, television, and radio should be analyzed for the following: Features Trends in content, stylistic features Creators or causes of content Content intentions Audience or effects of content Focus on attention, values, or "cultural patterns" of social groups
Cartographic material	- Diversified, although in general include maps with political and administrative division, hydrographic, ethnographic, ecologic, indication of crops and of land occupation modes, etc.
Publications	- Analysis of books, theses, dissertations, monographs, periodicals, research, etc. They undergo identification, location, compilation, and registration phases

analyses followed the postulations of Lakatos and Marconi (1991) and have already been notified by Medeiros (2016). We have pointed out usual terms regarding this specific field of scientific research and now we will present a rereading of these descriptions through a condensed proposition.

In view of the numerous possibilities of documental types and sources available, we realize that to better understand the research theme to be chosen, we must search and read related works, which happen to be in close harmony with the subject in discussion. This enlarging procedure together with placing different documental types in relation will help us find evidence on which to base our own conclusions and arguments in rejecting or accepting such documents throughout the data analysis process.

This analytical breadth-of-memory procedure in place is understood to be the handling of the document production context. In this perspective, among several aspects which the researcher must be aware of, one of them is to find out the authorship of the document. In addition, it is extremely important to critically research the social, cultural, political and economic context, the existence of any specific funding, as well as be conscious about the philosophical and scientific trends at the time of his writing.

Upon analyzing in a broad way research trends in this field, we can say that there have been three general patterns for the development of the investigations.

One of these tendencies would be the *Past-Contemporary Dialogical Study* (Box. 5.2), which is marked by the dialogic analysis between sources of different historical moments. Thus, articulations and approximations between the information collected will be carried out. It is important to note that the document's temporality imposes a time horizon for the dialogue between the consulted sources, whether past or contemporary.

Yet another trend observed is the establishment of the *Study of the History of the Element of Nature* (Box 5.2). The term "the element of nature" here refers to species, landscapes and cultural objects involved in nature's constitution. In this perspective, information is collected going back to the time the source of a determined element of nature was in focus, to other records produced so far over time. In the construction of the ethnobiologic narrative, the information obtained from both historical sources and

Box 5.2 Presentation of the investigative paths for research in historical ethnobiology based on Medeiros (2009).

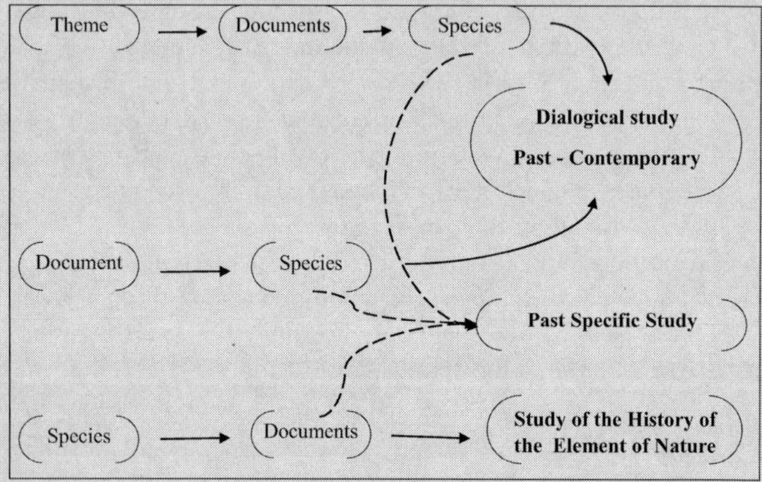

contemporary sources can contribute to a better undertanding of certain elements in nature through a more comprehensive view.

In another proposal, one can still develop research that is integrally focused on the historiographic analysis of the relation of people to elements in nature at a specific historical moment. It would be the *Past Specific Study* (Box 5.2). In this case, there is no concern for establishing a dialogue with sources encompassing the insertion of a certain element of nature at moments before and after that of the key source or the generating theme. The investigation ceases in the source dating period or occurrence of a historical fact.

Regardless of the perspective adopted, it is important to maintain the position analysis of the sources as diachronic so as not to risk making inferences about past events with the framework of cultural traces of the present century, which, in its turn, would characterize an anachronistic analysis.

*Therefore, we must seek to construct a scientific narrative in which interpretations can reveal the understandings of the authors expressed in the sources consulted and, as a result, ensure that our personal traces do will not dictate a truth different from what life was like in the past. This analytical procedure would lead to the construction of a discourse that frames information passed according to the present man's gaze, disregarding and disrespecting the cultural and temporal traces of the era of the document now under analysis.

2. Guideline for documental analysis in historical ethnobiology

Historical ethnobiology has to do with the principle of complexity. As such, there is no simplification as to how conduction can be summarized. This is to say that thinking about the methodological procedure pursuant to this ethnobiologic bias will lead us to deal with techniques utilized in several disciplinary areas. The specificity of each study will point the researcher to particular areas and techniques to be used in carrying out the research. However, from what has already been presented in this chapter, it is in order the establishment of some goals in summary form to help us find the right way in the methodological process, which is proper to historical ethnobiology. Thus, we will comment on the fundamental elements of this approach to condense the previously mentioned concepts and duly present a road map. The considerations we will present bellow include terms and ideas that we consider as being structural for developing a documental analysis. Thus, we will follow these previous concepts that we have kept in mind, and we will divide this guideline in six phases according to Dodebei (1997), and following Medeiros (2016).

Phase 1: Constitution of a coherent and homogeneous *corpus* of documents

This *corpus* is composed of documental material that expresses the thought of authors/social actors dedicated to the theme that is being focused on the research.

Phase 2: Establishment of a principle of choice

It encompasses the sampling definition that indicates the documental set. From the gathered documental universe, the most representative elements are thus selected and the research will be constituted of that significative information.

Phase 3: Definition of the analysis level and the elements in the material under study

At this stage of the research, it is necessary to define the significant elements for the documentary set under analysis. This way, it is possible to highlight references, explicit or not, informational traces (institutions, practices, and events). It is also possible to highlight the words used, paying attention to their grammar and their semantic usage from them, as well as the formal structure of the discourse and how they are considered in the context as a whole.

The informative content analysis is the way by which the universe of theoretical discourse analysis is performed, also called plan of ideas, as contained in the selected documents. The universe of theoretical discourse analysis encompasses three segments, which are the situational analysis of discourse, the theoretical discourse, and narrative. Firstly, as form of informative content analysis, the situational analysis of discourse comprises the text production considering its context at the time when it was produced. Whereas the second mode of analysis, the theoretical discourse, derives from an abstraction of the context creating autonomous articulations and illations. Finally, the narrative, which organizes through the discourse the sequence of events that have occurred (Kobashi, 1989).

Phase 4: Specification of an analysis method

The three segments that compose the universe of theoretical discourse analysis commented in the previous phase can be worked out from some methods of analysis. The need to specify these methods such as interpretive deciphering or analysis of frequencies and distributions will determine aspects related to quantitative analysis of the information collected from the text decomposition.

It is important to emphasize that the analysis of informative content establishes a sistematic classification in order to obtain correlations between the traces of the discourse seeking a conceptual comparison of the data collected. In this process, the textual differences observed are highlighted. It is then based on this type of analysis that a scale of complexity (non-singular qualitative traces of information) and specificity (singular qualitative traces of information) in which the categories of concepts are created.

Phase 5: Delimitation of the sets and subsets of the material under study

The action of establishing the goups and subgroups is a phase that gives us condition to propose articulations between the documents that compose the documental set under study. For this determination, descriptive elements are taken into account such as its geography (space, localities, and regions), temporal positioning (historical periods), social actors, biologic groups (botanic and zoologic), and minerals that are present in the document narrative. At this stage of the research, it is important and interesting to contemplate qualitative traces of the aspects under analysis into a broader context, including to consider the epistemological study and the origin of the informational traces under analysis. This movement of analysis is intended to check continuities (permanences) or non-continuities (changes) of these conceptual meannings.

Phase 6: Determination of the relations that characterize the documental elements

This determination of the relations will work with other perspective, which will be the understanding of a set of documents and their attributes (propositions, themes, or elements of nature), which will be analyzed. In addition, there is also the intention to provide a logical and causal relation propositions between the registers studied.

The basis of this procedure of working with the informational attributes of documents consists of the movement of observing the significative elements of the discourse, of verifying the relations between this information contained in the sources, and also of identifying the information regarding the individuals, the spaces and the period.

An essential procedure to be undertaken, when relating the attributes, it is not to lose the notion of the context in which the document under analysis was inserted. To achieve this broad view, the researcher should have as a basic principle the reading of the document itself. It should also be considered of fundamental relevance that this set of information reflects the constitution of the identity of social memory, implicitlly or explicitlly, which is presented in the origin and maintenance of the sources.

3. The constitution of scientific collections valued as biocultural heritage

Changing the angle of our discourse so far focused on the analysis of documental sources deposited in institutions for safeguard with the perspective to form documental source through the constitution of a biocultural collection, some key points must be made (Part 1) to build a path for the constitution of a biocultural collection (Part 2). Let us take a look at these points next.

3.1 Part 1. Forming documental source for biocultural collections

Considering the process of collecting and forming a biocultural collection lead us to think about the perspective of natural history museums, botanical gardens, and other institutions of botanic, zoologic, and humanistic knowledge. From this point of view, we must be able to identify in a given culture which are the representative objects present in the situations of everyday life, as well as those with a particular meaning or a historical meaning for that culture. Therefore, whatever is interesting to be collected for the formation of a biocultural collection could cover, for instance, objects utilized for eating, construction, and ritual practices.

Thinking about the formation of a biocultural collection, we can say that in the stage of collecting the objects that will be part of the biocultural collection, the ethnobiologist must be attentive to the registration of information that will tells us about all the moments of production of this biocultural element. In this documentation process, this observation includes knowledge, thecniques, remembrances tied to some natural material there will be used to produce a utilitarian object, which will have details and significance of a given culture. This documental procedure must cover notes such as persons involved, transportation, valuation, means of production, etc., in order to guarantee a comprehensive record of all the stages of processing the material that will integrate the biocultural collection.

The collection structure may be more universal or specific. A universal structure will have as a documentary fund representative samples of the material culture of different peoples, covering various historical periods and themes, depending on the proposal that the curator has in mind for such a biocultural collection. As a result, the structure of the more specific collection will be concentrated on objects that play a particular role in a particular culture. These objects with specific utilization refer to an activity (such as artifacts used in fruit harvesting, for hunting, cooking, cleaning, clothing, etc.), as well as to history or stories, which bear profound meanings in human cultures.

Considering these principles commented above, for the collection of objects, it is necessary to determine what is thought to be representative or culturally significant in terms of the production of material culture, so that an object can be selected to compose our biocultural collection. Thus, the selected material gains another status, which becomes a document that contains traces of memory from the place of origin. Regarding this role of producing the material in a biocultural collection, we have to collect as much information as possible about the object. This includes collecting the material used in its production, information about its production and uses, such as who is involved in the production, how it is used, what it is for, the local denomination, the history of the object and other aspects that allow for the greatest detail possible on the object. Thus, it will gain documental value, although it is outside its context of origin, that is, in another space (collection, exhibition, museums, etc.).

As for the space where the documented material must be deposited, we can say that the objects transformed into document can include, for example, collections of museums, scientific collections (botanical, zoological, or ethnographic), and educational collections (museum education, scientific education, etc.), even private collections. Depending on the place of memory in which the document objects are safeguarded and the vocation of the collection that this material will integrate, they will exercise various functions. The objects may be in the technical reserve of a museum, in permanent exhibition, serving a local exhibition, or undergoing actions by the handling of people at varying ages involved in educational and scientific actions. These multiple possibilities of deposit locations and object utilization purposes have a direct implication on the planning to determine what will be collected and their functions in the new axis of existence of the document object. Regardless of what the collected material is intended for, the principle of collection that should guide the researcher is to have a copy (and/or reproductions thereof) of quality that represents the culture of a community.

3.2 Part 2. Directional paths to biocultural collections

In the documentation of objects that is related to material cultural heritage, it is essential that every step of the processing of the raw materials that constitute the objects be recorded, along with how they were obtained and treated, the techniques and the instruments used in their extraction, preparation, and transportation, the social actors involved, and the final product characterization.

Documentation can be done in the form of a record written in a field book, for example, and in the form of image records, with photographs, scanning of images, videos shootings, and audio-recorded narratives.

Biocultural collections are integrated with specimens and other materials which includes, for example, objects or written notes. Considering specimens, we must not forget to collect samples of them. In addition, examples of raw materials (e.g., flowers, leaves, and tree barks) and different levels of processed materials must also be documented. So, next, we will make comments on the material documentation in the perspective of the formation of a biocultural collection.

3.2.1 Documenting objects for biocultural collections

The process of transforming a biocultural object into a document encompasses three plans that will be discussed. The first one is to document the object manufacturing process, which requires a recording of all the steps for its elaboration. Care must be taken to capture all the details in relation to the appearance and presentation of the object in every step of production. Another aspect that needs a lot of attention on the part of the researcher is the technique applied to the raw material, and in each part of the fabrication of a final object. To be successful in this endeavor, we can use photography, filming, drawing, sketches, and written notes in a field book, and any other type of visual records.

Another plant is the documentation on the aspects regarding the "know-how to do" of the object. This is important to understand the object in its entirety. Therefore, information on who participates in the elaboration of the object, how old they are, and what they use it for, for example, must be recorded in writing, as well as what strata of society is involved in the process of making and using it and how one can find out whether there are restrictions on use both in relation to gender and period. In this process, there should be a record of the details surrounding the social implications of the object related to a specific human society.

The final plan is to document cultural traces present in the existence of objects. To achieve this proposal, one measure is the undertaking of interviews with interlocutors who can reveal details regarding the in-depth meanings of local narratives. So, the focus in this perspective is to record information on the objects, so that we can learn about their use in a transcendental dimension (magical-religious uses), their importance for their culture, what would be the emic concept of manifestation of beauty directed to these objects, and how they have developed in the history of local culture.

3.2.2 *Visual communication of objects for biocultural collections*

The imagery record is characterized as an action that permeates the process of material documentation production. Photography, filming and any visual expression must be performed in a way intended to obtain a reliable record of each step in the making of the object, and also to attain records on how it is locally utilized. In this sense, the interviews will be fundamental for the apprehension of reports and images of people will tell us about their knowledge and practices of life and their relations with the registered objects. In this way, we will have a complete register of both people and objects in a community.

To attain an imagery record with scientific quality, some special care must be taken, which is possible through the use of varied thechniques, and if necessary, to go deeper into the art of photography or other visual communication to obtain goods results. Then we can create a great database for images and as well as for composing different works possible, such as books, articles, and also calls to action in the communities involved in the project with a view to learn how objects were used from natural resources.

3.2.3 *Collecting samples of plants, animals, and other elements that integrate objects for biocultural collections*

Species involved in the constitution of documented objects should have their samples collected and processed according to the standards of each specific area, for example botany, zoology, etc. After proper processing, the samples should be identified and incorporated into the collections of scientific collections, where they will receive a registration number according to their specific collection.

3.2.4 *Creating a documentary fund for biocultural collections*

The data set in hand will be diversified and composed basically of information, object and biologic samples deposited in collections, abiotic factors data (if pertinent), and imagery resources (photographs, filming, drawings, etc.). In view of this documental load, the researcher must organize all these items periodically during the research period. A crucial point in this organization of documental typologies, which are so variously different, is the fact that the relation between the items should not be lost in regard to the same object.

To maintain this unity between the documental sources of the same object, we can make use of appropriate references, which could be the register number in the scientific collection, the digital file names generated for specific information, or other instrument that promotes a systematization of the multiple supports of information regarding the object in a biocultural collection.

In general terms, we can think of two documentary funds of documental ordering. These will be composed of files that are related to the scientific collection (specimen data, object data, localization data, visual registers, and register number) and ethnobiologic data (including personal data of interviewees, interview forms, written consent, audio and visual registers). These two documental funds will be related and kept in unity through the appropriate collection numbers (registration number) of any associated species.

In this way, the data produced can be kept as a unit in a standardized way and accessed both by direct contact and virtually. This systematization practice will make the information accessible and related to each other at any time according to research needs.

We conclude this organizational dynamic with comments on the formation of documental sources, in case the starting point of the research is the current documentation on the dynamics between people and the elements of nature. Therefore, the researcher is enabled to elect elements of the material culture and elevate them to the condition of documental evidence to favor historical ethnobiology.

4. Basic materials required

For optimum research practice in historical ethnobiology, the researcher should take special care of some registration items and protection by organizing them before starting the field phase.

Registration items are viewed as a way to operate data collection and storage to make access to documents and personal notes possible at any time according to research needs.

These protection items provide the researcher with a barrier against the direct contact with waste, which may be present in the documents under consultation during the course of the project. Furthermore, these protective items guarantee a lower deterioration rate of the material to be consulted and, therefore, a more adequate conservation of them.

Table 5.6 Description of basic material registration and protection for the development of research in historical ethnobiology.

Basic material for registration	
Type of material	**Type of information**
Field note book	- In-depth, personal notes that seek and give meaning to the data that is being worked on
Reminder booklet	- Brief information that may arise in conversations and readings, such as the indication of a document to be consulted or the location of certain important material for the research
Bibliographic tabs and others or end note	- Descriptive data required for the location and composition of the document reference and bibliography consulted
Collection worksheet	- Content of documents broken down in full and in a way defined by the researcher, in specific fields of information. It may be adjusted to the nature of the data as the field work is processed
Notepad or field binder	The way in which research material should be compiled and preserved
Handheld scanner	—
Photographic camera	—
Hand magnifying glass	—
Basic material for protection	
Protective goggles	—
Disposable masks	—
Disposable gloves	—

In general, the registration and protection items can be consulted in Table 5.6. Certainly, the items described do not contain all the possibilities and needs for the work at hand. In this way, other materials can emerge and become essential to the research, depending on the project design, the nature of the documental mass and the information with which to work.

5. Closing words

The aspects discussed in this chapter constitute the methodological framework for the establishment of a research work in historical ethnobiology. The aspects involving first-order questions for the establishment of the research were dealt with, in regard to locations and places where the preserved vestiges are safeguarded, as well as the necessary methodological procedures, which are based on the biography document, and on the documental analysis.

Although the investigative view is trained on the past, and the encounter between the researcher and the traces analyzed, historical ethnobiology has been developed based on contemporary challenges. One of these noteworthy challenges is the narrative built by the investigation, which triggers proposals and reflections for the construction of a new reality, one which involves a certain aspect of the relationship of people with nature.

It is also in place to mention the challenges researchers meet when dealing with historical ethnobiology, namely, the demands from public policies that dictate how the models should be available for the collections. This issue brings up the fact that we may be dealing with objects that people of certain ethnicities or social status are not allowed to contact, for example. Thus, considering different reliefs that touch the existence of objects in a collection, it is important to understand biocultural collections as places for cross-cultural interaction, memory, and life. If we are to think along these lines, a public policy that determines the online availability of the collections can hurt the belief system held by the culture that produces the object here taken as a token for reflection. Furthermore, the ideal proposition would be the creation of an information system for biocultural collection. In this format, a policy of dialogue should be endeavored with the members of the various ethnic groups who produce the pieces, which are part of the collections. From this process, a dialogue would indicate which objects could be exhibited both physically and virtually.

Yet another challenge and requirement emerge with the heritage listing procedure. There is a lack of a system to guarantee heritage recognition. Our claim is based on the fact that we are well aware that it is only through these objects that we can learn about events and characters of the past. In this view, heritage recognition is justified in that it guarantees registration and memory preservation of the relation between people and the elements of nature in a certain historical moment.

With this closing reflection, let us move on to the next chapter to close our ideas on the subjects covered by historical ethnobiology.

References

Anon., 1777. The Modern Traveler. Trudy Lowndes, London.

Bellotto, H.L., 2006. Arquivos Permanentes: Tratamento Documental. Editora FGV, Rio de Janeiro.

Cassares, N.C., 2007. Acondicionamento de documentos e fotografias. III Curso Organização de Arquivos de Congregações Religiosas e I Seminário de Arquivos (Arqui)Diocesanos, São Paulo.
Chateaubriand, A., 1917. Carta autógrafa assinada. s.ed., Rio de Janeiro.
CONSELHO INTERNACIONAL DE ARQUIVOS., 2001. ISAD (G): Norma geral internacional de descrição arquivística, adotada pelo Comitê de Normas de Descrição, Estocolmo, Suécia, 19-22 de setembro de 1999, 2nd. Arquivo Nacional, Rio de Janeiro.
CONSELHO INTERNACIONAL DE ARQUIVOS., 2004. ISAAR (CPF): Norma internacional de registro de autoridade arquivística para entidades coletivas, pessoas e famílias, adotada pelo Comitê de Normas de Descrição, Canberra, Austrália, 27-30 de outubro de 2003, versão final aprovada pelo CIA, 2nd. Arquivo Nacional, Rio de Janeiro.
CONSELHO NACIONAL DE ARQUIVOS (BRASIL)., 2006. NOBRADE: Norma brasileira de descrição arquivística, 1st. Arquivo Nacional, Rio de Janeiro.
Dodebei, V. (1997). O sentido e o significado de documento para a memória social. Tese de doutorado, UFRJ, Rio de Janeiro.
Gastaminza, F. del V., 2005. La fotografia como objeto desde la perspectiva del análises documental. In: Aguayo, F., Roca, L., Coords (Eds.), Imágenes e investigación social. Instituto Mora, México.
Houaiss, A., et al., 1990. Normas Técnicas Para Transcrição e Edição de Documentos Manuscritos. Comissão de Sistematização e Redação do I Encontro Nacional de Normatização Paleográfica, São Paulo.
Khoury, Y.A., 2004. Guia dos arquivos das Santas Casas de Misericórdia do Brasil (fundadas entre 1500 e 1900). Imprensa Oficial do Estado de São Paulo/PUC-SP/CEDIC/FAPESP, São Paulo.
Kobashi, N.Y., 1989. Análise documentária e tipologias discursivas. In: Kobashi, N.Y. (Ed.), Análise documentária: considerações teóricas e experimentações. FEBAP, São Paulo.
Lakatos, E.M., Marconi, M.A., 1991. Fundamentos de metodologia científica. Atlas, São Paulo.
Le Goff, J., 1994. Documento/monumento. In: Le Goff, J. (Ed.), Memória e História. Editora da Unicamp, Campinas.
Lima, S.F. de, 2007. Organização de acervo fotográfico (fotografia em acervos históricos). III Curso Organização de Arquivos de Congregações Religiosas e I Seminário de Arquivos (Arqui)Diocesanos, São Paulo.
Lopes, E.M.T., Galvão, A.M.O., 2001. História da Educação. DP&A, Rio de Janeiro.
Medeiros, M.F.T., 2009. Etnobotânica histórica: princípios e procedimentos. NUPEEA/Sociedade Brasileira de Etnobiologia e Etnoecologia, Recife.
Medeiros, M.F.T., 2016. Historical Ethnobiology. In: Albuquerque, U.P.; Alves, R.R. (Eds.), Introduction to Ethnobiology. Springer, Switzerland, pp. 19–24.
Quivy, R., Campenhoudt, L., 2003. Manual de investigação em ciências sociais. Gradiva, Lisboa.
Rousso, H., 1996. A memória não é mais o que era. In: Ferreira, M. de M., Amado, J. (Eds.), Usos e abusos da história oral. Fundação Getúlio Vargas, Rio de Janeiro.
Werneck, A.L.L., Berwanger, A.R., Bacellar, C. de A.P., Alves, G., Leal, J.E.F., Marques, J., Bogaciovas, M.M.A., Flexor, M.H.O., Fonseca, V.M.M. da, Lima, Y.D., 1993. Normas técnicas para transcrição e edição de documentos manuscritos. Available from: www.portalan.arquivonacional.gov.br/Media/Transcreve.pdf.

CHAPTER SIX

General reflections on ethnobiology and education

Contents

1. Speaking about ethnobiology and education

From the perspective of science, we observe that thought and action are so naturally related to each other, that we are not even aware of the interference they have in human competences and their potential.

In this context, ethnobiology in its ethical, social, and ideological commitments toward science, society, and education, breaks with the vertical relations found in these, forming a scientific thought that is shared by other natural ethnosciences that do not conform to the complex movement from the center to the sociocultural and economic peripheries.

Ethnobiology in its interaction with the educational field has as proposal to safeguard all the essential values that permeate human culture, thus promoting a dialogue between the types of knowledge (formal and popular) specially during scientific education. On these occasions of integral formation of individuals, this science goes beyond the immediacy, that is, it reaches universality. At the same time, ethnobiology has a cultural objective, insofar

Historical Ethnobiology
ISBN: 978-0-12-816245-3
https://doi.org/10.1016/B978-0-12-816245-3.00006-8

as educational environments, and in particular the university, are the agents of culture. It is therefore visible that it plays an important role in the education process.

Thinking on the prism of education, it is fundamental to bring to reflection the meaning of the term. The term "educate" comes from the Latin *educare*, linked to *educere*, a verb composed by the prefix *ex* (that means out) + *ducere* (that means to conduct). Thus, "educate" is an act that conduct to out and insert the individuals into their own society.

Education can occur in a systematic, intentional form, organized in educational environments (formal education) or in a nonsystematic form beyond the proper spaces and specific time organized for that (informal education).

In this way, we perceive that it is quite common to overlap in understanding the meaning of the terms education and teaching. But, between the acts of educating and teaching, there is complementarity and distinction. Educating includes the orientation that allows a person to be integrated into his/her cultural group, while teaching includes an intentional and planned aid in the learning of a content.

Understanding the concept of education and also the concept of teaching, a question does not stop coming to mind and needs to be asked: why do education exists? Or, why do human societies educate?

The act of thinking about the systematization, organization and intentionality of the educational process occurred throughout human history. The aim was to establish a system in which cultural traits were passed on to younger generations. Thus, through this planning, the older generations were ensuring that the teaching of the chosen elements occurred as fundamental to the adaptation of the youngest to their social group. Thus, what we perceive is that in the teaching and learning process there is an intention, which is expressed by the marks of identity, ensuring that they are disseminated and maintained in their culture.

Following what was stipulated in the UNESCO report, the International Commission on Education for the 21st century (Delors et al., 1998), four basic foundations for education are defined. These comprise the following: 1. learn to know, "combining a general culture [...] with "learn to learn", to be able to take advantage of the possibilities that education offers throughout life"; 2. learn to do, "to acquire not only a professional qualification, but [...] [also] skills that enable people to deal with many situations and for teamwork"; 3. learn to live together, "developing the understanding of the other and the perception of the forms of interdependence, to carry out common projects and to be prepared to deal with conflicts, respecting values of pluralism, understanding, and peace"; 4. learn to be, "so that one's own personality flourishes better as well as the identity processes and that the person be able to work with a growing capacity for autonomy, judgment, and personal responsibility [making decisions], without underestimating any of the individual possibilities in education: memory, reasoning, aesthetic sense, physical abilities, ability to communicate" (Delors et al., 1998, p. 97).

With regard to the development of individual identity, the UNESCO report highlights the importance of allowing, in the first place, that each person be considered as a member of a particular local community. Thus, this identification provides the means for these people to consider other realities existing in other communities (Innovemos, 2008). These basic foundations encourage students to develop their critical thinking and to become open to other ways of conceiving life, of building thought.

Considering the teaching of ethnobiology, both at school or university, it is noticed that there is a growth in this interdisciplinary area. It is equally noted a greater number of activities and researches devoted to this study in the crossing field of ethnobiology with education.

Despite this scenario, some issues can be seen regarding these activities that involve researches and teaching practices related to the conjugation between scientific and popular knowledge. When we are in classrooms, the cultural knowledge of the students is naturally diverse from scientific knowledge. Thus, in this context, what matters is the teaching sciences marked by the demarcation action (Cobern and Loving, 2001), which presents itself as an attitude contrary to the regular annulment of cultural knowledge of the learner's life history.

The knowledge built through people's daily practice for their survival was managed through direct or indirect contact with the resources they had, which were relegated to the condition of second category of knowledge, due to the hegemony of scientific knowledge and the adoption of the center-periphery model, that verticalizes the understanding of others in their values, knowledge, and practice. In general, this knowledge from empiric origin is classified as being traditional or local (Medeiros and Albuquerque, 2012). Its valuation is reduced faced with the knowledge produced by Western science due to this secular process of domination.

In contraposition to this historical relegation to the condition of second-order knowledge, Baptista et al. (2014) point out that traditional knowledge does not need legitimation or recognition by science. Like scientific knowledge, those of a traditional or local nature follow their internal logic, as well as their validation, and applicability, with no need for external legitimation for those who utilize and perpetuate them.

Bringing this discussion to the scope of teaching, El-Hani and Mortimer (2007) dialogue about the science teaching under the perspective of a culturally sensitive scientific education and they affirm that the multiplicity of knowledge is preponderant for the creation of this model of education. We can bring this consideration to think about the teaching of ethnobiology, which can be built with this articulation between knowledge. For a greater immersion in our statements in this regard, we recommend reading the next chapter to access some important considerations about the types of knowledge and human communities.

2. Possible relations between ethnobiology and education

In the *curriculum* structure, science teaching that opens up to the inclusion and considers the traditional or local knowledge as an integral and important part of its *curriculum* has been discussed by some approaches. Among these ways of thinking of a *curriculum* from the perspective of transversality of knowledge, we can consider as main approaches the reflections generated by multiculturalism, universalism, and epistemological pluralism (Cobern and Loving, 2001; van Eijck and Roth, 2007). Let us see a little about each of these currents of thought.

In the current of multiculturalism, a multicultural scientific education is prioritized; that is, an education that takes place where colonizing and colonized cultures coexist (van Eijck and Roth, 2007). In this multicultural educational environment, it is believed that the fundamental ambience for scientific advancement can be created through the increase in the fields of relations where there is a multiple dialogue between the social actors and their knowledge. However, multiculturalists do not postulate that only the multiplicity of perspectives will be the unique source to activate multicultural knowledge within the teaching-learning process (Stanley and Brickhouse, 1994).

The argument of multiculturalists is exactly opposed to an attitude of exclusion, and it is therefore directed toward the inclusion of a multiplicity of knowledge in the structure of scientific *curricula*. There is also the understanding that a broader concept of science is necessary (El-Hani and Mortimer, 2007).

Putting themselves as a counterpoint approach to universalism, multiculturalists characterize universalists as being promoters of an exclusion policy. The point of questioning about the universalist discourse is based on the idea that there are several nuances that must be considered about lines of thought and a range of techniques that characterize science as not being so universal (Irzik, 2001).

It is precisely on this issue that universalism is placed, and it is characterized by defining science as being composed of universal knowledge. There is no room for the consideration about multiculturalism, although there is a recognition that other forms of knowledge bring contributions to science (El-Hani and Mortimer, 2007). However, it is certain that universalists do not believe that the multiculturalism would be feasible to be incorporated into scientific *curricula* and, following this line of thought, traces such as culture, gender, ethnicity, or sexual orientation of the cognoscent subject are considered worthless for the constitution of scientific knowledge (Stanley and Brickhouse, 1994).

In the third current, the epistemological pluralism, it is believed that any manifestation of knowledge must be considered and, thus, be part of the scientific *curricula*. The great difference of this line in relation to multiculturalism is that these varied forms of knowledge do not have the *sine qua non* condition to be viewed as a science (Baptista, 2015). The unfolding of this idea is that even traditional or local knowledge is part of the *curriculum*, in epistemological pluralism, scientific postulations will always be universal. Assuming this understanding, the other types of knowledge begin to be evaluated within their own cognitive systems, revealing the importance of acquiring these experiences (Cobern and Loving, 2001). With that, it is believed that these other nonscientific knowledges can perform a multiplicity of understandings that are seen as essential within the process of scientific education.

In this way, in epistemological pluralism, the *curriculum* must have its content being developed by docents in order to reinforce the most varied forms of knowledge. With this teaching practice procedure, Cobern and Loving (2001) affirm that what is intended in this current of thought is that the students develop knowledge of how science can have an advance when considering other forms of knowledge. Students will also be able to perceive the existence of other epistemologies that also unravel ideas that resemble scientific ideas. Finally, through this opening to other domains of knowledge, students will also perceive that, on the other hand, there is knowledge that was produced by science that is not possible to be constituted by other domains of knowledge.

Based on these ways of thinking about a scientific teaching that is inclusive in relation to different forms of knowledge, let us reflect on about the possible intersections between ethnobiology and education.

In a general way, we can say that in the works that have been trying to deal with this theme that contemplates both ethnobiology and education, one of the treated approaches turns to the study of the interactions, which encompasse several nuances observed in traditional knowledge, as well as scientific knowledge. In this investigative perspective, the urgency of understanding how the teaching-learning process can be considered in relation to the interventional experiences of students' traditional knowledge in the context of the school environment is brought to discussion. In this research path, we can notice that there is a set of thoughts that seeks to unravel these relations each one with its own assumptions, and appearing at different moments in the history of ethnobiology. A first group of works is characterized by the ideas that emerged from the movement of alternative conceptions. After this movement, the ideas that emerged gave way to what was the model of conceptual profiles, and afterward, multiculturalism brings its contributions.

There is another focus on the researches on this theme of ethnobiology and education. In these researches, the school environment is perceived as *locus* of investigation for understanding beyond the place in which this teaching environment is inserted. Thus, the school serves as a field of study for the development of different inquiry objectives in ethnobiology, such as the analysis of the students' conception about certain epistemological clippings.

Considering these two existing guidelines, we can perceive the existence of congruent aspects between them. A common feature is the existence of a comprehensive treatment of the interfaces between ethnobiology and education (e.g., Vougioukalou et al., 2014). However, the common trace can also be considered as only some of the elements present in the teaching-learning process, when then the focus of the research is restricted, for example, to the *curriculum* itself, or to the students (Baptista, 2012; Chandra, 2014; Kim and Dionne, 2014).

Another recurrent aspect in researches is related to the time of immersion in the school environment. The short duration of researches and the focus on the data collection whether by direct observation or the application of questionnaires are opposed to those in which the school community serves as a starting point for the development of pedagogical practices, with the creation of basis for didactic units, which really can bring changes to local communities (Baptista and El-Hani, 2007; Valderrama-Pérez et al., 2015). This process of joint and contextualized construction of a didactic sequence that integrates the different types of knowledge for science teaching is valuable insofar it promotes an education that becomes multicultural and that gains meaning with the community.

Thus, in the teaching process, it is necessary to think about a subject (student) in its entirety. It is used as a means of building this awareness and knowledge, the establishment of pedagogical practices that precisely promote this dialogue between different cultures and knowledge. Besides that, the conception of that varied scientific disciplines can contribute to the integral formation of the subject, promoting the expansion of interfaces between biology, humanities, history, exact areas, and so on. And why is the formation of the subject in a more holistic conception important? Because it is not only to live anyway, but to exist in plenitude (Viertler, 2006).

Therefore, when we pursue a holistic path of knowledge, we take into account the elements or principles that can provide us patterns that may indicate the clues to a deeper experience of transdisciplinarity. This transdisciplinary communication would cause, as it were, an innovation in the promotion of multicultural communication. The promotion of a dialogue zone would allow the various epistemological constitutions, belonging to each discipline, to reflect on their similarities and differences in conceptual and methodological terms (Viertler, 2006). The transdisciplinary communication is the central vehicle in promoting the scientific-cultural innovation process. It is the core for students, teachers, and the local community involved in the multicultural education process to have meaningful learnings.

This dialogical network can be seen in some cases beyond the teaching of science and biology as well as in ethnobiology. This is the case of the anthropologist Claude Lévi-Strauss who, to develop his structuralism, sought in the interface with linguistics the bases for his thought. Another anthropologist, Gregory Bateson, brought to his ethnography about Indonesian populations specific engineering concepts, such as the steady state. So too, the Greek mythology provided elements for Jung and Freud's reflective act on the human psyche (Viertler, 2006). These dialogues evidence how they are possible between different disciplines and how they create a creative sphere that propels scientific knowledge to go ahead in its development.

In addition to this scientific gain, it is increasingly important to rethink our way of bringing disciplines, knowledge and cultures together, and that it be used in favor of a multicultural education in which it is intended to promote democracy, respect for human rights, and a generation of a more embracing knowledge. The formation of subjects in their entirety, that is, that they represent more "a whole", without, however, losing their local references, can generate greater respect for cultural groups and promote a more symmetrical relation between them.

As we saw earlier, there are some currents of thought in the study of the relation between ethnobiology and education. We will look more deeply at some of these approaches that bring us to reflection of scientific education that considers other forms of knowledge. After describing some notes in this sense, let us also see some reflections around aspects involved in the interface between ethnobiology and education that will be relevant for the constitution of a teaching conception in ethnobiology.

2.1 Principles of the possible relations between ethnobiology and education: the movement of alternative conceptions

Regarding the alternative conceptions movement (ACM), we can situate in the 1970s the beginning of a movement about ideas developed by Piaget in the field of understanding the mental structures that lead to the construction of knowledge, especially regarding the students' previous conceptions that they brought to the formal educational environment (de Oliveira, 2005; Leão and Kalhil, 2015). Thus, in this sequence, a series of researches appeared that brought criticism to the Piaget's constructive theory, which at the time was at the beginning of its structuring. We notice that from these reflections and, specially, from the work of Driver and Easley in 1978, the ACM was formed (Mortimer, 1996).

The set of researches that made reflexive criticisms to Piaget's constructivism based its main arguments against the process of knowledge construction being decontextualized in the classroom and, also, in the fact that the pedagogical practice does not consider the symbolic dimension in the educational context, focusing only on the physical elements (Driver et al., 1999). Despite this contrary position, there was an agreement with constructivism with regard to the concept that learning is built by the subject, that is, the person actively makes knowledge flourish. In this context of the activity of the knowing subject, we can trace two important constitutive traces present in the ACM. One of them would be that the student plays a preponderant role in the development of knowledge, and the second trace would be based on the of students' previous ideas that also are of major importance in the development of education (Mortimer, 1996).

Considering these constitutive traces that are in the core of the ACM, we can perceive that this movement could be understood as being a mark for researches that wanted to have their focus on the understanding about a students' reality and their daily knowledge to gain strength in the academic environment. Thus, the ACM also brought a contribution of great importance to studies in ethnobiology. These researches absorbed the influences of the ACM and began to look for what possible connections with the area of education would be.

2.2 Other possible connections between ethnobiology and education: model of conceptual profile and multiculturalism

The model of conceptual profile began and was expressive between the 1980s and 1990s. The researchers who leveraged this model at that time postulated that for learning to be carried out it would be necessary for scientific concepts to follow a path composed of four moments. These would be as follows: 1. understanding of the concepts, that is, intelligibility; 2. solution of issues that were not solved by the knowledge already existing, that is, pulsability, to have vigor, vitality for their resolution, and thus, 3. fertility, that is, to be fertile, capable of resolving issues that were not yet resolved; and finally, 4. dissatisfaction, that is, to consider the previous conception unattractive, which could, however, become a conceptual conception more satisfactory than the previous one (El-Hanni, 2001). When going through these moments, the next moment would be to replace the students' previous knowledge with the new concepts that were learned. The educational process would therefore be centered on a student's individual cognition. With this cut, the model of conceptual profile model would lead us to the understanding that it is, to a certain extent, influenced by Piaget's studies on personal constructivism. What we can see in this pedagogical current is that there is no focus of the educational process on interaction and affectivity, thus, learning is devoid of significance and sensitivity beyond formal context.

In this prism, the model of conceptual profile works with worldviews. One of these views is contained in scientific concepts with its own codes. What students have in them of previous cultural knowledge comprises another set of worldviews. In the face of these distinct ways of seeing and formulating conceptualizations about the existent world, a state of conflict is generated during the learning process due to the fact that students bring new worldviews that confront scientific concepts (El-Hani, 2001). The centrality of this concept of worldviews in contextual constructivism places social influences and cultural identity as fundamental aspects for the learning process (Cobern, 1994).

Two learning movements are distinguished by the model of conceptual profile. One of them occurs when the student goes through the understanding, but not through the apprehension of scientific knowledge. In this case, there was a modification of the student's conceptual profile (Mortimer, 1995). Another situation occurs when there is the comprehension and, also, the apprehension of the knowledge. In this case, a conceptual exchange takes place. This is a process in which there is a movement of cognitive exchange, when the old knowledge is substituted by the new scientific knowledge (El-Hani, 2001).

Considering this dialogue between scientific concepts and worldview during the realization of the act of teaching, ethnobiology could be inserted in this context considering the contribution that it could bring to science teaching. Working with students' worldviews, ethnobiology would incorporate into science teaching concepts taken as formal, aspects that are from within the students, such as their forms of classification, numeric notions, and also notions of time and space (El-Hani, 2001). In this way, a greater understanding of the conceptual conflicts between teachers and students could be promoted, a better understanding of existing forms of science considering their specific contexts, in addition to the creation of a sphere of valorization of the different forms of culture knowledge formation.

Assuming this form of multicultural teaching, teachers would promote during the educational action a favorable environment to respect and intercultural conversation between the most diverse worldviews and academic science. In this type of teaching environment, the practiced scientific education would be based on a larger concept of science. Following this path, science would be the incorporation of concepts from different cultural origins, which would, therefore, distance it from the universal concept and legitimacy conferred on Western science (Mortimer and El-Hani, 2007).

It would be precisely in this mediation between different concepts of science that the conceptual profile would act. It would be a means by which understanding between the most varied domains of knowledge could be achieved, establishing parallels of similarities and distances, nuances, limitations, comprehensions, in short, establishing a critical analysis of the peculiar characteristics of each way of knowing nature. And in this sense, the use of methods and techniques of ethnobiology could benefit this promotion of integration between what was built of knowledge following academic science and what students bring to classroom, which are their individual stories and elements of a collective memory that is also specific and related to where they come from. Ethnobiology in science education would promote the acceptance of different universes, as well as stimulate interaction between the actors involded in the educational process (students among themselves, teachers with students), as well as fostering mutual respect between each modality of knowledge and its actors.

Thus, the teacher who works with the conjugation between teaching and ethnobiology for the establishment of a multicultural dialogue between the scientific concepts existing in a classroom has as pedagogical practice to highlight the students' universes, and the same way, to bring to students the theoretical contribution built throughout the history of science.

2.3 On educational environments and integration of traditional knowledge through the relation between ethnobiology and education

Considering the educational environment as a possibility for the development of an ethnobiologic reflection, a variety of approaches can be proposed. We believe that ethnobiologic research in this perspective gains space and it should address the perception of actors involved in education regarding scientific themes, such as analysis of the environments or any specific characteristc of elements of nature.

From researches that are developed with individuals from educational communities can emmerge a variety of memory traces regarding traditional or local knowledge present in the localities where the education units are inserted or from where their students and/or teachers came. This truth represents the necessary understanding for the acceptance of researches of this bias within ethnobiology because through these, it is possible to advance in a better notion of the existing relations between people and their living environment.

Besides that, when conducting an ethnobiologic research in an educational environment, there is a possibility of a greater dialogue between the forms of knowledge during the investigation process, which includes the stages of data collection and return activities. This dialogical characteristic of ethnobiologic research highlights the mediating role of the ethnobiologist who gives importance to the dialogue in the pedagogical practice.

In this sense, it is necessary to consider some aspects of the integration of the traditional or local knowledge through the relation between ethnobiology and education. Ethnobiology can represent an indispensable science, in the measure that it can be used by researchers and educators in its applicability related to cultural groups and nature conservation.

Traditional or local knowledge has played a crucial role in the history of the search for the survival of the most diverse cultural groups. These knowledge systems have given these societies autonomy and the ability to adapt to situations that have arisen over time. As they are systems of a dynamic nature, the knowledge of local communities encompasses both the traits fixed in individual and collective memory, as well as those that are added to this existing knowledge. However, with the process of modernization and dissolution of the old community life, people ended up distancing themselves from nature and from this set of knowledge, practices and values present in the local memory that gave them the possibility to self-supply or manage. It also puts at risk the existence of knowledge that ends up losing meaning due to the new living conditions of people, more distant from nature in their daily life. The weakness of this management and maintenance skill affects, for example, food security and the production of other goods essential to people's survival (López and Rodríguez, 2014).

A possible path for the conservation of knowledge from traditional or local systems occurs through educational action. To achieve this proposal of conservation, scientific research is not the only one that has value. It is also necessary to integrate communities, especially the children and youth of these cultural groups, through education. Thus, the conceptual set constituted by ethnobiology and education is the right way to favor the perfect integration between the cultural heritage associated with biodiversity and biocultural preservation. Therefore, it becomes of utmost relevance to think how education can consider the knowledge of communities as part of its *curriculum*, just as it is necessary to mobilize teachers to be sensitive to multicultural scientific education.

2.4 Formal and informal education and the incorporation of traditional knowledge

An important aspect in the scope of reflections on the interfaces between ethnobiology and education is the integration of traditional or local knowledge within *curricula*. Next, we will see how this integration would take place within formal education.

According to López and Rodríguez (2014), in a multiplicity of contexts of the formal education, from early childhood education to undergraduate or graduate education, it would be fundamental to guarantee that these local contents will be part of the *curriculum*, which means that they will compose the formal education. This action could be carried out through the choice of a set of basic contents through which traditional or local knowledge would be integrated into the *curriculum*.

Thus, according to this structural thought, there would be the selection of ethnobiologic contents and their application in the *curriculum*. The first task would be to assemble a team of ethnobiologist researchers who would lead a working group to establish the elements belonging to traditional knowledge because their importance as a content. It would be essential to integrate and promote dialogue between different educators, that means, ethnobiologists, formal and informal teachers. When finalized the research on traditional or local knowledge, the contents could be integrated as theme in different academic disciplines, such as natural sciences and arts.

From this point, the proposal that was articulated could be used in teaching practice, following two reflections related to the educational environment. Firstly, we could present the students' relation with the school environment where the educational situations take place. In this space, students integrate themselves with different notions and types of knowledge (formal and informal contents), since traces of local cultural expression, individual and social memories live together. In this reflection, taking Baptista (2010), classrooms should to be seen as multicultural spaces, and teachers should know the previous conceptions of everyday life brought by students, in order to supply knowledge with this mixture of realities (internal society, school, and external societies). And then, we could present life outside of school, when we can observe that ethnobiology is versatile and can be developed through activities both in the educational environment and in the informal educational environment, such as the student's community environment. Considering this integrative didatic practice as a way of developing a content, what we can perceive is that students are presented with conditions in which they can form a conscience about the importance of traditional or local knowledge that reinforces their identities. In addition, there is an opportunity for students to realize the importance of knowledge from mature generations in their communities. In this case, ethnobiology places itself as mediator, facilitating a combination of social actors of different age groups.

Now, we present a proposal for a pedagogical practice that could be applied in education for this integration to occur. A good didactic sequence could be structured based on some pedagogically planned meetings. We suggest a sequence that would include an order starting from a theoretical class (an example could be a class where there would be an exposition of theoretical aspects, where an informal interaction between students and teachers would be made, discussing the contents in a common didatic form, including comments about the species and, to end the activity, it may be use a thechnological resource such as images, videos, sounds, etc.); an external class (identification and collection of samples of species in the field work for practical activities); a practical exercise (done with the samples collected in the field work) and, for closing the content, a round of discussion about the results achieved by each student.

2.5 Qualification of educators

Another aspect that must be reflected on is the excellence of teaching in scientific training carried out within the scope of university teaching environments and in the continued assistance to students in their academic formation. This formal environment in which training takes place must excel in solidity, in the sense of promoting scientific research and stimulating teachers to have a critical consciousness of the issues involved in the teaching-learning process devoted to ethnobiology. Within this theme of qualification of educators, we can mention that there has been a systematic implementation of disciplines at the university at different levels with regard to ethnobiology in some localities. These educational actions allow trainning teachers and scientists, in addition to promoting the insertion of ethnobiology in the issues that arise throughout history, meeting global necessities and also sustainability, as well as the wishes of local communities.

In view of this reality, an increasing number of professionals with autonomy is necessary to the better development of educational practices in ethnobiology. We need teachers who feel comfortable in implementing new techniques and processes. We also need training for university professors, and for that, postgraduation is essential for creating the conditions for a formative experience in education. And, furthermore, it is necessary to build a university identity, in which formal education is not only the training of professionals, but also comprise a circle of science and culture at a high level. It is this circulation of ideas that we need for implementing good teacher training.

After talking about the necessity of postgraduate, let us move on to the question of the need to demarcate specific disciplines related to ethnobiology in undergraduates. These should be based on solid objectives related both to the formal scientific and cultural dimensions. The binomial formed by the association between science education and ethnobiology can be conceived as a permanent and dynamic movement. The teaching of ethnobiology has generally been addressed through extension courses, debate cycles, activities at academic meetings, and as modules within the *curricula* of different levels of undergraduate training. There is a differentiated form of organization of disciplinary systemization that normally composes the institutional teaching *curriculum*. In this way, we can observe the insertion of ethnobiology in it and speaking particularly about disciplines in which ethnobotany is a part of its content, we find it as a theme in disciplines such as economic botany, medicinal plants, general botany or human ecology. We believe that this way of thinking about the curricular structure of courses in which ethnobiology could be inserted is a reflection of a prejudiced devaluation of ethnobiology because, to a large extent, it has been considered as a theme within established disciplines or as elective disciplines, but its recognition as an area of knowledge is gradually expanding.

What we would like to emphasize is that the initiatives commented above should be increased with the desire to broadly structure the *curriculum* of ethnobiology, thinking about which courses and subjects must be offered, and also reflecting on pedagogical strategies and didatic methods and materials. Within this perspective, a right path would be to place the *curriculum* followed by many institutions in a place of debate for its reflection. We talk here about the *curriculum* since a great part of the issues related to teaching are linked to the *curriculum*. It is the *curriculum* structure that is configured as the motive power that models formal education. As D'Ambrósio (2001) preconizes, the *curriculum* should be taken as an educational pedagogical planning. Thus, the pedagogical planning must be rethought and evaluated from time to time, allowing the reflections that are posed as current issues in society to be considered in the act of educating within the teaching institutions.

There is a range of topics within ethnobiology that could be dialogued with students. It is suggested that the teachers seek for an integration between these topics by valuing both inter- and transdisciplinarity. Adopting this stance, the teachers could provide their students with a broader reflection on topics, such as those related to the sustainability of biodiversity (conservation and management). The planning of the pedagogical action could include discussions about the significance and practice of the use of elements of nature in the local communities. It is also interesting to keep in mind that during these discussions students can perceive different biases arround the perception of the social actors in the region studied. Thus, the elaboration of strategic plans for the use and conservation of flora and fauna is a practical activity that they can execute with the orientation of the teacher.

In the teaching of ethnobiology, the approach to these topics, could create the conditions for good discussions on other biases, such as the regulation on access to traditional knowledge and possibilities of benefit sharing, which are evoked from the social contexts both global or local.

Considering what has been said, it is clear that the path of inclusion and dialogue between knowledge is configured as a possibility for the curricular components to propose an integration based on similarities and differences between different modalities of knowledge, such as the traditional/local, the scientific, as well as the academic knowledge in the formal institutions of education during the exercise of teaching practice.

This broad understanding of forms of knowledge will create an identity for the teachers who consider and are sensitive to intercultural dialogue with their future students, creating a teaching awareness that the exercise of their teaching practice will involve the act of placing themselves as mediators between daily knowledge of their students and the scientific concepts that will form the content that they will be dealing when they teach in the future.

The establishment of connections, similarities between traditional/local, scientific and academic knowledge as an integral part of the contents of the curricular components courses, will contribute to the training of teachers who will seek to develop teaching strategies that incorporate the dialogue between distinct cultures. In addition to this aspect, this training that brings a content that respects and values other forms of knowledge about nature will enable students to broaden their understanding of what nature is under different prisms. What will be achieved will be a scientific education sensitive to cultural diversity, that values students' daily knowledge and that, in this way, they are encouraged to understand the scientific concepts.

2.6 Scientific education sensitive to cultural multiplicity

Speaking specifically about Brazil, we realize the existence of several cultural groups, which have a proper system of knowledge that guides their relations with their living environment (Macêdo and Baptista, 2014). These cultural groups, the so-called traditional communities are represented, for instance, in Brazil, by indigenous peoples, Azoreans, Amazonians, besiegers, artisanal fishermen, cowboys, "ribeirinhos" (riverside people), "campeiros" (peasants), and "quilombolas" (black groups) (Diegues and Arruda, 2001).

Continuing to take the Brazilian reality as an example, even though there are traditional communities that are found all over the national territory, school and university education in Brazil continues with a curricular inclusion in which the knowledge about nature coming from these communities is seldom addressed. In an educational environment, these traditional communities are often present through their social actors who enter the formal education system (Baptista, 2010). However, teaching opportunities are rare in which the teachers include in their content everyday knowledge and social memory of their students, so that there is an integration of this traditional knowledge with that scientific-academic knowledge through allowing a multicultural character in the adopted pedagogy (Baptista, 2010).

Although there are legal instruments promulgated by Brazilian public policy, with documents that guide a new education in Brazil (*Parâmetros Curriculares Nacionais, PCNs* [National *Curriculum* Parameters], Brazil, 2006), this renewal has not been finding concreteness in most educational institutes. In the new parameters of the Brazilian public education policy, an important point stands out that is given to the students' previous experiences, adding them to the learning of knowledge worked in the context of formal education. For the teaching of biology, the field of study of this discipline is highlighted considering the human being and the mediation of culture in its relations with nature and the life of nature itself. Nevertheless, teaching persists in being scientific, reproducing the idea of a unique and universal scientific culture, based on modern Western science, which is formed by academic scientific knowledge. What continues to happen is an education in which this academic knowledge is directed to the students of biology disregarding what their identity really represents. Teaching has therefore been detached from the structural reality of traditional communities, becoming a decontextualized educational practice (Macêdo and Baptista, 2014). In this process, learning becomes deficient to the extent that the understanding of concepts distances and disregards what the students bring in themselves as experiences.

Weaving a few more reflections within a general context about this scientific teaching, we would like to comment that its reproduction has been occurring not only by a single explanatory fact. Instead, we could mention some factors that contribute to this practice. We consider that among these factors, teacher training is crucial. A preparation of the new teacher that is done by promoting the idea of cultural diversity in teaching can create a new possible path. It is about breaking an academic structure in which universalist conceptions of the human being are perpetuated in terms of their history, knowledge, science and other competences regarding teaching training (Arroyo, 2008). What we are thinking about is a teacher training that deals with cultural diversity, which is sensitive to what students bring to the moment when teaching takes place, but not that it is based exclusively on the transmission of knowledge by the teacher.

As another way, the teaching of science and biology, which is based on the valorization of dialogue between different cultures, is seen as a possibility of inclusive education. In this bias, teaching must prioritize the contextualization of the contents. Thinking like this, it is believed that there will be a better opportunity for students to understand scientific concepts through cognitive processes of similarity and difference when relating these concepts with those that they already bring in themselves as memory devices.

Within a global context, bringing traditional communities back to our reflections, as they have a hundred knowledges about the environments in which they are inserted, multicultural teaching can be a means by which it becomes viable for establishment of these relations of similarity and distances between the scientific knowledge of the academy and traditional knowledge of the students and their communities. It is precisely trough this intercultural dialogue established and stimulated during the process of the teaching-learning process that it is possible to contribute to a change in teaching (Macêdo and Baptista, 2014). That is, scientific education will have the possibility of being much more interactive if we move concretely toward scientific education sensitive to cultural diversity.

In this procedure that values the contextualized and multicultural education, in addition to the preponderant role of dialogue for the significant learning of scientific content, we can also emphasize the importance of considering epistemological pluralism, multicultural scientific education, conceptual constructivism, and the niches of students' worldview during the planning and concretization of an educative action aimed at learning, in fact, charged with meaning for students. (Baptista and El-Hani, 2007).

In this bias, it becomes essential to create the condition for deconstructing prejudices about different forms of knowledge with students. Thus, this understanding would be an unfolding of science teaching in which different modes of knowledge would be unravel to students through the pedagogical situations presented during the learning process. Students themselves would come to the conclusions of how and to what extent other modes of knowledge interact with scientific making. This procedure of demarcation of knowledge during formal education events allows students to have an approximation about the existence of different epistemologies (Cobern and Loving, 2001). There is a space to realize that there are multiple ways to seek understanding of elements of nature, landscapes, abiotic factors, etc., which ends up not depositing in Western science the credential of the single, absolute, and therefore, universal truth.

Thus, what we must consider is the value that each culture has in the world. There is no comparative space to say whether this knowledge is better than another, whether from marginal cultures or those considered central. What matters is that each culture has its value immersed in its history, social memory, and social ties which perpetuate its beliefs and identity signs throughout history. Clifford Geertz (1999) brings in his ideas a critical stance on the asymmetrical consideration that can be observed among knowledge and proposes that an anthropology of thought be made in search of more symmetry. Thus, it is necessary to emphasize and accept that traditional knowledge has an experimental character, thinking of biodiversity differently from academic knowledge, it is were, it has an understanding of nature based on its life built through daily practices.

From the point of view of teachers training who value cultural diversity, we can say that this way of thinking about teaching with the inclusion of traditional knowledge in their practice, it is to work with the feeling of teachers so that they can perceive the meaning of the cultural diversity of their students. The teaching process should be guided by this openness and sensitivity, as a means to improve the quality of learning (Baptista, 2006). It becomes a process in which teachers will devote their planning time to prepare contextualized teaching strategies that reflect a critical construction of scientific knowledge, also encompassing traditional knowledge. Thus, teachers will build a grid harmonizing these strands through the choices of information and methods coherent with this perspective. It is in this way of proceeding that we find the action of teacher training who, in the future already as teachers, will assume this this form of conceiving education based on integration between disciplines, knowledge and cultural traditions.

In this respect, Darrel Addin Posey (1987) presents a methodology for ethnobiology in which it would promote the dialogue with respect between the academy and the cosmology of individuals from traditional communities. We can even make a reflective cutout on the use of this methodology in the teaching-learning process with the students with whom we establish a formal educational relation.

In this proposal for multicultural education, we must not disregard the central role of interaction that is promoted by dialogicity.This concept would be at the core of the educational process, and should also be seen as an exercise of the right of freedom for all individuals involved during educational training (Freire, 1981). Therefore, we could also try to achieve greater symmetry in the integration related to varied types of epistemological knowledge and methodologies, as advocated by Bruno Latour (1983, 2000). However, in addition to this symmetry in the dialogue, in which teachers seek to know and value the culture of their students, another aspect should also be considered. Another important factor in the teaching-learning process comprises didactic resources. These elements assist in learning and should be thought of as available and useful tools that can have a great power to promote interaction between scientific knowledge and other fields of knowledge (Baptista, 2007). Using the methodological tools of ethnobiology, the teachers could, for example, make an investigation with their students to get closer to the knowledge that is typical of their local realities. Then, a set of teaching materials could be created together with the students, so there was care to adapt these materials with traditional knowledge of the students themselves In this way, teaching material resources would have an effectiveness as a means of facilitating the teaching-learning process of formal contents of scientific teaching.

Among the confrontations present in the search for the exercise of a scientific education attentive to cultural diversity are the training of teachers; the mediation of interpersonal conflicts, that is, the conflicts that arise from varied perceptions of the world within the classroom; the openness of students in valuing traditional knowledge inside and outside their school environment; the concretization of a learning based on contextual constructivism, in which students would maintain their daily knowledge and develop their cognitive *spectrum* with the incorporation of scientific knowledge; and, finally, a confrontation would be to create the conditions for students to have the possibility to know and effectively choose the best use of traditional and scientific knowledge for their daily lives (Baptista and El-Hanni, 2007). Despite the multiplicity of aspects involved in the effort so we have a science concept that aggregates other forms of knowledge of nature, the pedagogical gain is invaluable. Reflection on new meanings opens up to teaching practice that prioritizes interaction and integration between multiple cultures (Baptista and Carvalho, 2015). Therefore, this expansion of the spectrum of the students acts in the direction of creating space for them to perceive all the nuances, meanings, feelings and contexts, that is, the whole range of knowledge, whether in the teaching environment or through the daily knowledge considered during learning.

3. Conclusive words

After considering some of the contours that compound the complex relation between ethnobiology and education, we can close this chapter with the reflection that thinking about scientific education within ethnobiology is to consider a pedagogical practice that will be thought, concretized, evaluated and restructured in each real situation in which the educational process occurs.

In this sense, what we can say with certainty is that education in ethnobiology is inextricably linked to the idea of constant teacher training and attentive reflection to achieve a teaching practice in which, in each new situation, its pedagogical action can provide an environment, a methodology and a dialogue with students who consider the local reality. This productive instability means not framing the teaching of ethnobiology concepts within a unique pedagogical line, which is general and universal. It is not a matter of proposing an exclusive path to understand, reflect and discuss about it. On the contrary, this constant state of circulation of concepts and practices indicates that we must observe how education acts in ethnobiology in order to develop a specific educational culture for ethnobiology. This pedagogical practice only thought by and for ethnobiology can become concrete, either through teaching practices exercised by teachers or through the institutions that make room for this science in the training of human resources.

What we seek is to understand education in ethnobiology emphasizing the absence of rigidity in its pedagogical practices. In this way, what we intend to do would be to observe its upward curve in the educational process through new articulations and new didactic experiences within this process.

Thus, circulation and constant rethinking are essential components of those who propose themselves as an agent of education in ethnobiology. That is, this multicultural educational formation should not be understood as having a static character, immutable. Far from this idea of conceptual and practical stability, education in ethnobiology is always understood as having a mutable character, that is, it is in permanent alteration.

Therefore, with the intention of reviewing each experience and putting us in constant attention to think about the act of educating in ethnobiology, it is what enables us in this teaching process to capture its intrinsic characteristics tied to human beings with their practices and values between themselves and the nature. What occurs is that education in ethnobiology, due to its dynamic aspect, is historical because it happens and changes according to time, and therefore, each teaching situation is a new experience. We believe that this character of constant reflection and construction reflects the great purpose of educating in ehtnobiology, which is to encourage the development of identities sensitive to its issues and discussions. And concluding, as the identities of the apprentice subjects are always varied, their education cannot be considered equally universal and static, but dynamic *par excellence*.

References

Arroyo, M.G., 2008. Os coletivos diversos repolitizam formação. In: Diniz-Pereira, J.E., Leão, G., Org (Eds.), Quando a diversidade interroga a formação docente. Autêntica, Belo Horizonte, pp. 11–36.

Baptista, G.C.S., 2006. Conhecimentos prévios sobre a natureza, prática de ensino e formação docente em ciências. Revista da FAEEBA – Educação e Contemporaneidade 15 (26), 199–210.

Baptista, G.C.S., 2007. A contribuição da etnobiologia para o ensino e a aprendizagem de ciências: estudo de caso em uma escola pública do estado da Bahia. Universidade Federal da Bahia, Salvador; Universidade Estadual de Feira de Santana, Feira de Santana.

Baptista, G.C.S., 2010. Importância da demarcação de saberes no ensino de Ciências para sociedades tradicionais. Ciência & Educação 16 (3), 679–694.

Baptista, G.C.S., 2012. Elaboração de materiais didáticos como apoio ao diálogo entre saberes no ensino de biologia nas escolas do campo. Revista Iberoamericana de Educación 60 (4), 1–11.

Baptista, G.C.S., 2015. Um enfoque etnobiológico na formação do professor de ciências sensível à diversidade cultural: estudo de caso. Ciência & Educação 21 (3), 585–603.

Baptista, G.C.S., Carvalho, G.S., 2015. Science as a cultural activity: comparative study of Brazilian and Portuguese teachers' conceptions about science. Proc. Soc. Behav. Sci. 174, 69–76.

Baptista, G.C.S., El-Hani, C.N., 2007. Diálogo entre modos de conhecer no ensino de Biologia: estudo de caso numa escola pública do Estado da Bahia. Encontro Nascional de Pesquisa em Educação em Ciências. Anais, Florianópolis. http://www.nutes.ufrj.br/abrapec/vienpec/CR2/p200.pdf (Accessed 15.03.20).

Baptista, G.C.S., Vargas-Clavijo, M., Costa Neto, E.M., 2014. Introdução. In: Baptista, G.C.S., Vargas-Clavijo, M., Costa Neto, E.M., Orgs (Eds.), A etnobiologia na educação ibero-americana: compreensão holística e pluricultural da biologia. UEFS Editora, Feira de Santana, pp. 11–18.

BRASIL, 2006. Parâmetros Curriculares Nacionais: ciências da natureza, matemática e suas tecnologias. MEC, Brasília.

Chandra, D.V., 2014. Re-examining the importance of indigenous perspectives in the western environmental education for sustainability: "from tribal to mainstream education". J. Teach. Educ. Sustain. 16 (1), 117–127.

Cobern, W.W., 1994. World view, culture, and science education. Sci. Lit. Cult. Stud. Proj. 5 (4), 5–8.

Cobern, W.W., Loving, C.C., 2001. Defining "science" in a multicultural world: implications for science education. Sci. Educ. 85 (1), 50–67.

de Oliveira, S.S., 2005. Concepções alternativas e ensino de biologia: como utilizar estratégias diferenciadas na formação inicial de licenciados. Educar 26, 233–250.

Delors, J., Amagi, A.I., Carneiro, R., Chung, F., Geremek, B., Gorham, W., Kornhauser, A., Manley, M., Quero, M.P., Savané, M.-A., Singh, K., Stavenhagen, R., Suhr, M.W., Nanzhao, Z., 1998. Educação um tesouro a descobrir: relatório para a UNESCO da Comissão Internacional sobre Educação para o século XXI. Cortez, São Paulo; UNESCO, MEC, Brasília.

Diegues, A.C., Arruda, R.S.V., 2001. Saberes tradicionais e biodiversidade no Brasil. Ministério do Meio Ambiente. USP, São Paulo, Brasília.

Driver, R., Asoko, H., Leach, J., Mortimer, E., Scott, P., 1999. Construindo conhecimento científico na sala de aula. Química Nova na Escola 9, 31–40.

D'Ambrósio, U., 2001. Etnomatemática: elo entre a tradição e a modernidade. Autêntica, Belo Horizonte.

El-Hani, C.N., 2001. Porque a etnobiologia e a etnoecologia são importantes? uma resposta do ponto de vista da educação. Encontro Baiano de Etnobiologia e Etnoecologia. Universidade Estadual de Feira de Santana, Feira de Santana.

El-Hani, C.N., Mortimer, E.F., 2007. Multicultural education, pragmatism, and the goals of science teaching. Cult. Stud. Sci. Educ. 2, 657–702.

Freire, P., 1981. Ação cultural para a liberdade. Paz e Terra, Rio de Janeiro.

Geertz, C., 1999. Os usos da biodiversidade. Horizontes Antropol. 5 (10), 13–34.

Innovemos, 2008. Red Regional de Innovaciones Educativas para América Latina y El Caribe. Educación y diversidad Cultural: Lecciones desde La práctica innovadora em América Latina. Publicado por La Oficina Regional de Educación de La UNESCO para América Latina y El Caribe OREALC/UNESCO Santiago. http://unesdoc.unesco.org/images/0016/001626/162699s.pdf (Accessed 21.02.2020).

Irzik, G., 2001. Universalism, multiculturalism, and science education. Sci. Educ. 85 (1), 71–73.

Kim, E.-J.A., Dionne, L., 2014. Traditional ecological knowledge in science education and its integration in grades 7 and 8 Canadian science curriculum documents. Can. J. Sci. Math. Technol. Educ. 14 (4), 311–329.

Latour, B., 1983. Give me a laboratory and I will raise the world. In: Knorrcetina, K., Mullkay, M. (Eds.), Science Observed: Perspectives on the Social Study of Science. Sage, London, pp. 141–170.

Latour, B., 2000. Ciência em ação: como seguir cientistas e engenheiros sociedade afora. Editora UNESP, São Paulo.

Leão, N.M. de M., Kalhil, J.B., 2015. Concepções alternativas e os conceitos científicos: uma contribuição para o ensino de ciências. Latin-Am. J. Phys. Educ. 9 (4), 2–4.

López, A.V., Rodríguez, J.F., 2014. Reflexões sobre a floresta dentro e fora da escola. In: Baptista, G.C.S., Vargas-Clavijo, M., Costa Neto, E.M., Orgs (Eds.), A etnobiologia na educação ibero-americana: compreensão holística e pluricultural da biologia. UEFS Editora, Feira de Santana, pp. 387–409.

Macêdo, L.S.S.R., Baptista, G.C.S., 2014. A contribuição dos conhecimentos tradicionais para a formação docente em biologia sensível à diversidade cultural. In: Baptista, G.C.S., Vargas-Clavijo, M., Costa Neto, E.M., Orgs (Eds.), A etnobiologia na educação ibero-americana: compreensão holística e pluricultural da biologia. UEFS Editora, Feira de Santana, pp. 479–499.

Medeiros, M.F.T., Albuquerque de, U.P., 2012. Dicionário Brasileiro de Etnobiologia e Etnoecologia. SBEE/NUPEEA, Recife.

Mortimer, E.F., 1995. Conceptual change or conceptual profile change? Sci. Educ. 4, 267–285.
Mortimer, E.F., 1996. Construtivismo, mudança conceitual e ensino de ciência: para onde vamos? Investigações em Ensino de Ciências 1 (1), 20–39.
Posey, D.A., 1987. Etnobiologia: teoria e prática. In: Ribeiro, B.G., Org (Eds.), Suma Etnológica Brasileira, v. 1. Etnobiologia. Vozes, Petrópolis, pp. 15–25.
Stanley, W.B., Brickhouse, N.W., 1994. Multiculturalism, universalism, and science education. Sci. Educ. 78 (4), 387–398.
Valderrama-Pérez, D.F., Andrade, A.M., El-Hani, C.N., 2015. Dialogue between scientific and traditional knowledge in the science classroom: development study of a teaching sequence in a school in Taganga (Magdalena, Colombia). Proc. Soc. Behav. Sci. 167, 217–222.
van Eijck, M., Roth, W.-M., 2007. Keeping the local local: recalibrating the status of science and traditional ecological knowledge (TEK) in education. Sci. Educ. 91 (1), 926–947.
Viertler, R.B., 2006. Métodos antropológicos como ferramenta para estudos em etnobiologia e etnoecologia. In: Amoroso, M.C. de M., Ming, L.C., da Silva, S.M.P. (Eds.), Métodos de coleta e análise de dados em etnobiologia, etnoecologia e disciplinas correlatas. UNESP/CNPq, São Paulo, pp. 11–29.
Vougioukalou, S.A., Barfield, K., Huish, R.D., Shiels, L., Brosi, S.L., Harrison, P., 2014. The contribution of ethnobiology to teaching plant sciences: student and faculty perspectives. In: Quave, C.L. (Ed.), Innovative Strategies for Teaching in the Plant Sciences. Springer Science, New York, pp. 33–45.

CHAPTER SEVEN

Thinking about the conceptualizations of types of knowledge and human communities

Contents

1. Knowledge and culture

When we think about a teaching of ethnobiology that articulates different modalities of knowledge, we can not help but reflect on the concepts about communities and also about the knowledge produced by them throughout history. Therefore, we propose an immersion in this chapter on the types of knowledge and human communities that will bring us a broader vision that will help us better understand the interfaces between the fields of ethnobiology and education, before we go directly to the teaching of historical ethnobiology.

To talk about knowledge is to speak of a culture where the entire knowledge system takes place. This cultural system is shared variably by members of a human society. In this way, the culture of a society, a people, a social

Historical Ethnobiology
ISBN: 978-0-12-816245-3
https://doi.org/10.1016/B978-0-12-816245-3.00007-X

representation, as well as of an ethnic or religious group has inference in different dimensions of the existence of the peoples who are bearers of that culture. Among these unfoldings, we can mention the extensions of thought, and language, those of desires, feelings and meanings of the elements of the world, as also those related to social behavior, values, conceptions and expressions of creativity, among other possible dimensions that integrate human life.

Considering this initial understanding of knowledge and culture, it is important to say that we must be aware not to assume a common attitude that tends to means culture as to have something materially or immaterially acquired. Instead, we should turn to the understanding that culture is nothing other than an abstract comprehension of knowledge that is disseminated among individuals in their communities. In addition to this proposition, we must also remember that a culture is dynamic; that is, it is always in motion, always changing as it suffers new cultural or biologic and evolutionary pressures or even recurrent pressures that reach it. Although culture refers to knowledge distinctly distributed among individuals in a community, the sharing of meanings in the daily life of these same individuals is a socially lived process. It is not just about experiencing exclusively individual processes, but what occurs is an interaction between the individual memory that is inserted in a social group forming a collective memory. On the other hand, the collective memory also influences individual memory. Then, cultural and social structures place people conditioned to assume certain roles and to acquire social identities as a part of a whole, of a social group. Thus, the collective memory of a social and identity character is constituted.

Regarding the transmission of knowledge among individuals in community, we can say that it identifies with the acts of copy of information or also with the practice of teaching and learning social traces important to the survival. We can characterize it as being the process by which the knowledge, as well as the other dimensions of the social plot, such as language, perceptions of the world, values, and behaviors accepted by the community collectivity are transmitted among people who integrate a specific community. This process occurs from the interrelations between individuals from communities or even between these people and all material or immaterial conjectures that arise throughout the historical process of societies (Heyes, 1994; Boyd and Richerson, 2005; Hoppit and Laland, 2008; Mesoudi and Whitten, 2008). These placements lead us to the understanding that the transmission of knowledge can occur in a given culture through individuals of the same community, or equally, the transference can take place between individuals from different cultural groups. Furthermore, it is still possible to carry out a direct or indirect transfer of information in this social context of knowledge transmission.

Currently, a theoretical guide for studies on the transmission of knowledge has been the theory of cultural evolution (TCE). This theory assumes that the Modern Theory of Evolution is the path to understand cultural systems, since they are constituted by evolutionary properties (Cavalli-Sforza and Feldman, 1981; Mesoudi et al., 2004, 2006; Mesoudi, 2007).

The premises defended by TCE assume as inheritance of information both culture and genetic basis throughout the process of human evolution. Other basic concepts employed by TCE seek to understand the cultural dimension by explaining its diversity and development over time by different cultural groups. TCE then seeks to consider culture through methods used in biologic evolution studies (Mesoudi et al., 2006).

Thus, some of the traits that characterize TCE comprise aspects such as considering the dynamic aspect of cultural systems and the reaffirmation of a coevolutionary dynamic constituted by the relation of these cultural systems and the genetic system, as well as endorsing the possibility of acquiring nonparental characters (Cavalli-Sforza and Feldman, 1981; Mesoudi et al., 2004, 2006; Boyd and Richerson, 2005; Richerson and Boyd, 2005; Mesoudi, 2007).

In view of these guiding traits, studies on the transmission of knowledge based on this type of evolutionary approach, the so-called TCE, usually seek to answer some fundamental issues, which would be especially focused on the type of information and transmission period regarding its level of importance related to a given community; and consequently there is in these transmissions an aspect of choice that will determinate if a particular information would be chosen and desirable to be appreciated by some individuals; if there are singular ways to these transmissions occur; and if there would be anyone of the community to be elected as a model of influence to be followed by the others of the community. To this last aspect we must make an additional comment about the models of knowledge transmission in where we find: vertical transmission model (from parents to descendants); horizontal transmission (between individuals of a unique generation); oblique transmission (between generations, when young people have adults as reference, but they are not their parents); one-to-many transmission (from a leader, an audiovisual media, such as a book, magazine, digital or analogical vehicles of information, to many individuals in the cultural group); and, the model known as many-to-one or concerted (from the oldest individuals to the younger people of the cultural group) (Cavalli-Sforza and Feldman, 1981; Hewlett and Cavalli-Sforza, 1986).

It is clear to us that knowledge and culture go together, and multiple are the ways in which the transmission of knowledge takes place. After this immersion about the understanding of knowledge as a cultural element and the theoretical current that seeks to understand its transmission, we will approach the types of knowledge identified. To this end, we will divide our comments in two blocks, one dedicated to making some considerations from a methodology systematized by Trujillo Ferrari (1974), and another, expanding some of these concepts to other types of knowledge underlying one of the categories of knowledge presented by the author.

1.1 Methodology conceived by Trujillo Ferrari (1974) about the types of knowledge

In the approach of Trujillo Ferrari (1974) are presented some specific characteristics attributed to four types of knowledge defined by him, namely scientific knowledge, philosophical knowledge, religious knowledge, and popular knowledge, about which we find descriptive items in Table 7.1.

After presenting the characteristic points referring to the types of knowledge, we can now establish some parallels between what distinguishes each of them.

Philosophical knowledge is marked by the use of the pure reason, and only it, to make the propositions of the issues that involve the life of the

Table 7.1 Description of the types of knowledge based on Trujillo Ferrari (1974).

Type of knowledge	Characteristics	Descriptions
Scientific knowledge	Real (factual)	- Presence of occurences or real facts
	Contigent	- Presence of propositions or hypotheses based on experience in order to prove their veracity or falsity
	Systematic	- Presence of a theory through an ordered logic system
	Verifiable	- Exclusion of all statements (hypotheses) that can not be tested and proven
	Fallible	- Presence of non-definitive, or conclusive knowledge

Table 7.1 Description of the types of knowledge based on Trujillo Ferrari (1974).—cont'd

Type of knowledge	Characteristics	Descriptions
	Approximately exact	- Presence of new propositions and techniques that could restructure the theories that already circulates in this hall of knowledge
Philosophical knowledge	Attribution of values	- Presence of hypotheses based on experience and not observation
	Unverifiable	- Presence of hypotheses not available for refutation or confirmation
	Rational	- Presence of a group of statements logically connected each other
	Systematic	- Presence of coherent hypotheses and propositions through the coherent apprehension of reality
	Infallible and exact	- Presence of hypotheses and propositions based on the accuracy of the apprehension of reality
Religious knowledge (theological)	Attribution of values	- Presence of valued propositions based on sacred doctrines
	Inspirational	- Presence of doctrines inspired by the supernatural
	Infallible	- Presence of unquestionable truths based on the supernatural
	Systematic	- Presence of a systematic order provided by a supernatural creator
	Unverifiable	- Presence of a belief based on the supernatural not requiring verification
Popular knowledge	Strength of values	- Presence of a known object influenced by the values of the knowing subject

(Continued)

Table 7.1 Description of the types of knowledge based on Trujillo Ferrari (1974).—cont'd

Type of knowledge	Characteristics	Descriptions
	Reflective	-Presence of a reflection conditioned to the subject's intimacy with the object
	Unsystematic	- Presence of a systematization of ideas based on the subject's own experiences
	Verifiable	- Presence of a verifiable perception based on the subject's daily life
	Fallible and inexact	- Exclusion of hypotheses about general phenomena because it is based on the perceptions of the subject and the community

human being. Through reason, a value judment of what is right and wrong is developed during the problematization of the issues.

In relation to the object of study, scientific knowledge use techniques and direct or indirect sensory observations to analyse concrete facts, on the other hand, the philosophycal knowledge uses experiences to analyse conceptual and logical relations of ideas.

Regarding the method, science takes the experimental method as a way of maintaining its excellence, as it is centered on concrete facts that are set and validated by experimentation, but, on the contrary, philosophy that excels in the logic of ideas is based on the rational method using deductive process, where conceptual ideas are sistematyzed in a logical chain of reasoning (Ruiz, 1979).

With regard to the procedure, both types of knowledge also have their particular characteristics. In this way, the procedure adopted by science is directed to parts constitutive of the object under analysis, then, its focus is the understanding of fragments of reality. Philosophy, on the other hand, has in view general formulations that consider the universe through unified conceptions that value the universe as a unit.

Considering religious knowledge it gives truths an infallible condition because they represent in the supernatural in themselves. It is a matter of believing or not, where there is no doubt and then there is no concern to validate them.

About the concept of vulgar, popular knowledge (common sense), it is centered on the subjects' proximity to the objects and situations experienced, which is composed of a set of information that is ordered not always critically by the individuals, giving importance to feelings, sensitive knowledge, concrete visualizations, and through of an informative language made by the individuals themselves, as well as that made by others who reach them (Ander-Egg, 1978). Thus, vulgar or popular knowledge, in a broad way, refers to the subjects' knowledge that is current and naturally developed in their daily life in the face of the countless experiences that involve them in their environment. It is configured as the knowledge without a system or a method to obtain it (Babini, 1957). In comparison with scientific knowledge, this popular knowledge differs radically from the scientific knowledge in just one question: what is the way to acquire knowledge?

Both popular knowledge and science intend to acquire knowledge. Scientific knowledge is objective, impersonal and intend to understand reality by formulating hypotheses that can be verified with the aid of theories. In turn, popular knowledge has a characteristic that is circumscribed in reality by a subjective and personal formation of its concepts, bringing individuals closer to an ethnocentric position on how to produce, learn and circulate knowledge among individuals through their individual and collective memory that will influence their perceptions and practices.

1.1.1 Local knowledge

We also emphasize that, underlying popular knowledge, we find human groups that are different from the cultural point of view, and therefore, they manifest a specific, local knowledge, traditions, practices, beliefs, meanings and feelings expressed and maintained by these cultural groups over time (Warbunton and Martin, 1999; Vandebroek et al., 2011).

In this way, we can find a local knowledge taken as a synonym of traditional or indigenous knowledge, but this local knowledge does not only concern the origin inhabitants of a special place or even specific human groups. In addition, all communities, no matter if rural or urban, can develop and maintain this knowledge over the time. Finally, besides rural and urban communities, this term refers to communities that may be sedentary or not, original or migrants inhabitants of different regions of the world (Medeiros and Albuquerque, 2012).

The place, in this case, has a great influence on the development and systematization of the cognitive scheme. These, in turn, are related to culture, nature, and ecology (Miranda, 2005). This is due to the fact that human beings immerse in their environment seek to understand everything that permeate their relations looking for their deep significances. Thus, people express their informational heritage constructed along their experiences, which are maintained and preserved in the social local memory.

The term "local" also has some different nuances in academic interpretation. These controversies refer mainly to this term because it is understood that it does not express the real value of this knowledge in a broader context, disregarding the discussion of this knowledge on a global scale. However, this term gained more understanding as contexts at different scales began to be confronted, that is, a local scale with the most global scenarios (Nazarea, 1999).

The issues involved in the concept of local knowledge, regarding knowledge about natural systems, also deal with them from multiple perspectives. Among them, we can mention studies, for instance, on biodiversity conservation, folk taxonomies and others, which comprise ways of working with articulated conceptualization with themes treated in cognitive, epistemological, ethnobiologic, and anthropologic approaches (Escobar, 2005).

1.1.2 Traditional knowledge

Considering the nuances of knowledge, we also refer to traditional knowledge. This is produced and developed by the collectivety (community), having as a principle the wide dissemination and assimilation of concepts, knowledge, and information transmitted between individuals from one generation to another as a cultural heritage (Santilli, 2005).

Following this definition, traditional knowledge is not only related to information about organisms. Thus, this type of knowledge also encompasses the dimensions that are linked to human constructions of meanings about the environment and also to the interaction of this physical environment with elements of nature (Albuquerque, 2005).

Thus, traditional knowledge comprises a set of information, which is represented by all the manifestations of the people through the imaginary, elements of immaterial and material heritage, and beliefs according to particular dimensions of the world. This set of information, as we already pointed out, belongs to the collectivity in a given traditional culture, and its dissemination and perpetuation usually takes place through orality (Diegues and Arruda, 2001).

In relation to the term itself, "traditional" is related to a conception of knowledge stability. This form of interpretation of the meaning of the term related to knowledge systems imputes to the traditional knowledge proper to cultural groups a primitive character. This configuration of being crystallized in the past is, therefore, opposed to its real nature, which is to be in constant adaptation and in a creative dynamism during its process of construction and dissemination.

It is a conception of knowledge that is not concerned with formulating universal laws about nature. On the other hand, when comparing to scientific knowledge, traditional knowledge can also help to give answers to situations that integrate social and cultural realities of individuals in traditional communities that share and hold this knowledge (Medeiros and Albuquerque, 2012). We note that not only in their own societies can this knowledge be useful, but also in societies that exceed their limits of territorial occupation. Especially when we think in terms of conservation, the knowledge of those who preserve their traditional wisdoms can be invaluable sources of models for the use of nature.

1.1.3 Neotraditional knowledge

Neotraditional knowledge encompasses traditional knowledge together with new knowledge from outside a given community (Begossi, 1998; Berkes, 1999).

This new knowledge may have been acquired from the interaction between distinct cultural groups intentionally or even by an unconscious action (Medeiros and Albuquerque, 2012). This process of cultural approximation occurs when the assimilation of new information can be aggregate to the immaterial heritage passing to the condition of collective memory.

1.1.3.1 Indigenous knowledge

It is usual to find the designation of "indigenous knowledge" associated with traditional knowledge. However, it must be made clear that the indigenous denomination does not encompass other forms of societies, but indigenous. Thus, the application of "indigenous knowledge" is configured as an expression that is restricted to those who present themselves within this specific societal configuration. Talking a little more about this issue of the term "indigenous", it is often in tune with the idea of primitive, but what we need to enlight is that in reality this term is imbued with a strong ethnic significance that represents an important resistance of centuries. We also point out the knowledge belonging to indigenous culture, which, according to English anthropologist Posey (1947—2001) (1999), is a knowledge consisting of formative traits that compose a peculiar cultural system of indigenous matrix.

This type of knowledge can be conserved through a collective memory, that is, in the social scope of an indigenous community. It can also be conserved, for instance, by individuals who maintain individual memory in favor of the collectiveness depending on their social role in the community.

1.1.4 Concluding on the types of knowledge

What we can trace as a closing of this brief approach to popular knowledge is that this knowledge, with all its nuances, is knowledge constructed through the lives and experiences of peoples, whether indigenous, descendants of the African diaspora or rural communities, for example. Another characteristic is that this knowledge is passed on essentially through orality and from generation to generation. And, they are also essentially dynamic, based on culturally constructed values and beliefs, used in people's daily lives and also allow adaptation processes linked to the exclusion or incorporation of information to occur to adjust to specific situations.

Finally, we refer to a term called "ethno-knowledge", that has been frequently used in specialized literature, replacing the term "local knowledge" or "traditional knowledge", as noted by Alves and Albuquerque (2005). In this way, we can associate a general meaning for the term "ethno-knowledge", by affirming that it refers to knowledge produced by varied ethnicities and in different localities from popular knowledge or wisdom. This would be an expression that would have a broader character and, at the same type, exclude the possible skewed interpretations for all typologies encompassed by the knowledge that is produced by the "other".

It is also interesting to note that in the methodological scheme systematized by Trujillo Ferrari (1974), among scientific, philosophical, religious, and popular knowledge, the subject himself can apprehend the significance of objects in tune with various typologies. Thus, it is possible that the representativeness of the types of knowledge may be present in the same person, helping him to understand the most varied aspects of his existence. A final idea that we would like to state is that we must consider that all these types of knowledge have their importance and, therefore, we must consider them in a symmetrically.

1.2 Different perspectives of knowledge: emic and ethical

In view of the cultural identity traits of different populations, we verify how expressive are the coexistences that are established among the explanatory details that constitute the various cultures. This coexistence between multiple knowledges is being constructed among individuals from communities or from one and another social group throughout history. When we take into account that talking about knowledge is touching on what culture consists of, we wonder what would then be the relation between knowledge and cultures from the perspective of ethical and emic approaches.

Considering these ethical and emic approaches, we must put ourselves in a state of attention when analyzing the cultural patterns of a given community, of a society. The examination of other ways of life that are put in contact with our own cultural patterns, which is characterized as ethnocentrism, and thus, as an ethnocentric posture, lead us to achieve biased results. This bias skewed of the information perceived and analyzed about a given society ends up producing results that can only be partially mentioned. This is due to the fact that by performing a detailed analysis of the cultural signals of culture other than our own, we can see that the evidence of a specific culture is different from the cultural patterns with which we are intimate and which, therefore, we are familiar with, because they are part of our constitution, of our life experience.

Historically, the differentiated conceptualization between emic and ethical originates in the proposal of analogy of these terms with the linguistics categories: phonemic and phonetic (with the phonemic and phonetic linguistic categories) consolidated by the American linguist and anthropologist Kenneth Pike (1912–2000) (1954). In Pike's construction, emic would be part of ethical and the opposite would also be true. His approach is essentially synchronous. The starting point would be the verbal expression to reach the nonverbal (Pike, 1978). This signifies that the ethical is a way in which the emic is achieved. According to Pike (1990), a hermeneutic circle is formed by the emic and the ethical, which is always in continuity, and is not interrupted. To illustrate this dynamic between emic and ethical, Pike (1990, p.45) elaborates a singular poem, the "Emic Circle", which says: "See, and know. Know, and be. Be, and do. Do, and see. See, and know", thus demonstrating the chain that does not break in the emic-ethical complex.

During the 1960s, following a different line of thought, the American anthropologist Marvin Harris (1927–2001) (1964, 1969, 1976, 1990), who developed cultural materialism, elaborated a proposal to make an ethical anthropology. In this way of thinking, anthropology should disregard the narratives of social actors in relation to their own attitudes, and it would be developed unically through the observation of behavior. Opposing what had been proposed by Pike, Harris believed that the understanding of the structure of the emic would not be found in itself, but only though an outside perspective. Therefore, Harris understands that a cultural complex seen from an external point of view requeries an external linguistic manifestation, which is the ethical, to assist in the understanding of the emic.

The search for an emic-ethical distinction, already in the 1980s, found space beyond the anthropological field. Thus, several other areas of knowledge appropriated the emic-ethical during this period (Hussey, 1989).

Also, in the following decade of 1990, Harris himself will revisit his writings with the formulations of distinctions between emic and ethical and will recognize that both mental and behavioral classification would not be dialoguing with the emic and ethical by virtue of the way the conceptualizations had been worked on in his writings. At that time, the emic represented the interior of the "natives" and that could be observed through their external manifestations, such as speech, habits, customs, etc.

Thus, rethinking the emic-ethical complex, Harris reviews the anthropologic literature and recognizes that it is vast in exemplifying ethical discourses that sought to point the mechanism of the mental structure of the human being. Harris was referring to the examples in which the observer, that is, the anthropologist, defined the meaning of what was inside the mind of the observed, in this case, the "native". Otherwise, Harris will begin an understanding that the cognitive mechanism of the human being does not occur exclusively through a parallel between binary categories which is framed in an universal mental structure. That is, who determines the existence of these opposite binary categories, as example, "black - white", "hot - cold", was the observer. They were not necessarily realized by the observed. Reflecting on this, Harris then assumes that is feasible for mental processes to be contemplated both emically and ethically, as well as the behavior stream. What was put into reflection was that the observed phenomenon must be analyzed by means of a comparison between the emic-ethical both related to the mental structure and the behavior of the observed people.

In this sense, attention should be focused on perceiving the distinction between what people really think and what they say they think, so also with regard to what they say they do and what they actually do. This exercise of observation of the other is commented by Viertler (2002, p.23), when she affirms that "is necessary to be very cautious during the decoding of the real cultural meaning of verbalizations and behaviors of people". Since the cultural context is far from that of the researcher's origin, more care should be taken in this investigative exercise. What is sought to establish is then the articulations between "what is spoken, what is accomplished, and what is thought" (Viertler, 2002, p.23), that is, the connection between speech, action, and thought. The focus of the observation of distinction of these elements occurs since the subject can "think about something, but may not talk about it [...], and not act according to that outlined thought [; or the observed subject may] think about that something, talk about it, but nevertheless not act as he [thought and] spoke [; or, still, the subject may be able] to do something, but do not think or talk about it" (Viertler, 2002, p.23), in addition to the other possibilities of conjugation existing between thinking-speaking-acting.

Thus, what we can perceive is that what differentiates the emic and the ethical is the subject himself. Although the ethical approach is necessary, in the emic approach the subject is seen as being an integral element of a given community or society. In this way, this subject builds his perception from within this group. That is to say that the emic approach enlighten how the individuals culturally distinct understand their identity characteristics. Otherwise, in the ethical approach, the process of observation takes place from a person outside the observed social group. Therefore, in the exercise of ethical observation, there is an implicit approach that is carried out automatically. And in the emic approach, because it is a way of perceiving the world through the parameters of the other, it is necessary to do an exigent exercise, with bigger perseverance, expending greater energy, because it is not a movement of natural understanding of the observer (Rosa and Orey, 2012).

In this context of reflection, we can suggest that for the process of emic-ethical distinction during the observation of individuals in a community, the observer is in constant analogy between what seems strange to him and what sounds familiar to him, between the alternation of who is the observer, whether from inside or outside the community, between being a member of the community or not being, in addition to other possible specific analogies of researching the other (Campos, 2006).

The important thing is that in this process of observation of the other always reigns respect, whether in relation to a subject of the present or the past, which is accessed through the historical documentation that reaches our hands. Also do not include postures and expressions that demonstrate ethnocentrism to achieve symmetry in dealing with cultural traits that integrate the most varied forms of human thought (Latour, 1999).

After these considerations about the more general types of knowledge that compound human thought, let us move on to some comments that will lead us to immersion in the concept of the term "community", which until this time has been mentioned within a more concise weighting.

2. Communities

First of all, we will situate the term "community". *Communitate*, from Latin, has as meaning a diversity of concepts, depending on the *spectrum* that is tried to understand this term. The theorical approach used to achieve the understanding of the term "community" may vary. Common traits or great acquiescence of the community are the marks of these conceptions, endorsed by linguistic dictionaries.

Bringing to the social sciences the term "community" will find its sense in social relations. This means that there is a sharing of traits among the members that make up the community. In the meaning conferred by the social sciences, "community" usually involves the convivence of a group with particular characteristics and practices, and also, coexisting in the same space (territory).

The concept of community, by the social sciences, was realized at a time when there was a strong indication for the decline of this form of social organization that was until then was in force. This scenario of change and loss due to the installation of modernization and capitalism functioned as a promoter of the beginning of folklore studies. These were dedicated to group, catalog and analyze the material and immaterial elements of cultural heritage, such as symbols and discourses of popular culture. That is, the origin of studies on popular culture had its source in the belief that popular cultural traits would be lost; they would disappear in Europe (Burke, 1989).

Throughout the 19th century, social experience was linked to this context in which there was a dynamic of transformation of the values and social arrangements of the community by the implementation of a new structure based on modernity and capitalism (Lifschitz, 2011). This process of decline of the community in favor of this new social form will leave its historical mark in the writings of authors of theoretical and political subjects of different natures. Among these, we can mention by way of example the German sociologist Ferdinand Tönnies (1855–1936), the French sociologist anthropologist and philosopher David Émile Durkheim (1858–1917), and the Prussian sociologist, historian, and political philosopher Karl Marx (1818–1883), who imprinted in their discourse the idea that the decline and even the destruction of community arrangements were caused by the modernity and capitalism.

Taking Tönnies, among the cited authors, we will seek to deepen our concept of community from the construction presented by him. Before, however, let us say a few words about this author. Tönnies was born in

Germany, in a community with a well-enraized family structure named Frisian, a community of independent peasants with a long ancestry. Tönnies published *Community and Society* in 1881, a work that serves as a reference till to present days for the study of community, where he postulates a different proposition for the conception of the community in social sciences (Tönnies, 1947). His theoretical approach turned to a systematic treatment of the community theme. Thus, the concept and community studies with a theoretical approach can find in Tönnies the inscription of its origin. Considering this founding importance, we will expose data on the construction of Tönnies.

Community and Society had its first version published in 1881. It was subtitled *Empiric Forms of Culture*, which alluded to the epistemological controversy that at that time existed in Germany in the scientific sphere at that time. There was, therefore, a contrast between historicism and those who defended the Enlightenment rationalism and were positivists. On the one hand, there was historicism that advocated that the sciences of culture, despite being dedicated to the study of values and truths even in a relative scope, and in different historical and cultural scenarios, should be considered as sciences. Otherwise, positivists and Enlightenment rationalism claimed that these sicences of culture should be able to formulate universal laws such as those proposed by the so-called hard sciences in order to be considered as sciences. Tönnies' position in these two strands was to adhere to the historicist argumentation. However, he also argued that to formulate scientific theories about society, it would be essential to consider social categories in a broader analysis. Developing this line, he discussed two sociologies, one called pure sociology and another that would be applied sociology, wishing to inscribe the community in social theory as a category of universal and comprehensive scope.

Tönnies continues his theoretical development by employing references from the science of his time and proposes a conceptual analysis linking the community with the organic and the society with the mechanical. With this proposition, he moves toward an understanding of the community as being a body in where all its parts functions well in terms of physiologic representation and in the same way for the experiences lived by this body an existential representation. It is worth saying that the community excels for a unit because a complex of bodies are interrelated between themselves in a same territory, and then, invariably will constitute a unique body.

Building this *metaphor*, Tönnies establishes his argumentation as an "analytically necessary fiction" (Tönnies, 1995, p.233). Through the constitution of this metaphor, it was feasible to perceive the community as a unique body, but that at the same time it brings in it particular and harmonic characteristic, which are different from others communities.

The structure of the book *Community and Society* continues to be woven by Tönnies, and he describes in the first part of this book, in the chapter *Theory of the Community*, the actors of primary socialization covering the relations between mother and child, brothers, couples, and friends. In these relations of primary socialization, all types of affective and natural units are clearly observed. Community relations, the bonds that are formed with other people, would be the unfolding of this so-called natural affectivity. In this way, what was conceived was that there is a particular systematization between consanguinity, affectivity, and community, which was called the blood community.

Other community forms are also analyzed by Tönnies. One of these associations proposed by him is the "community of place", which was conceived in relation to the place (territory) in which a given cultural group of the same family or neighborhood lived.

These two circumscriptions of community as community of blood and place are the aspects most emphasized in the concept of community. Despite this, Tönnies still brings another contour to the understanding of the community. This would be relative to the community of the spirit, a concept that we can consider to be more differentiated or uncommon than the others already commented. The community of spirit has a immaterial form, based on intuition, subjectivity, and intimate values, but that also refers to social group and territory.

Following the terminology adopted by Tönnies the communities could be understood as being of blood, place, and spirit, which are reproduced within the borders of the most comprehensive society. Thus, the concept of community was conceived during the 19th century from the first reflections carried out in Europe, as a socialization established from a common origin, consanguinity, as well as from interior manifestations expressed by the proximity of the subjects.

Over time, passing 100 years, social transformations occurred and that community that served as source for Tönnies' observations and formulations had undergone changes. New observations were made by researchers for the sociological study, who observed that the limits or, as already pointed out in this same chapter, the borders were attenuating. One of these authors was Durkheim, who associated this *phenomenon* with the transformation of community through the intensification of trade and, consenquently, with the division of labor, and also with the increase of the population, which propitiated a great expansion in community life. But, returning to Tönnies' thinking, he thought differently and emphasized in his discourse the protagonism of external agents on the social change of communities, that means, an exogen concept of societary changes.

The fact is that in this community process of major changes, some communities have "dissolved" towards cities. However, as Maffesoli (1987) states, the sociability of a reduced group "insists" and communities are still alive. This validity of communities, or rather, their permanence, is linked to the fact that they bring in them the primary need of humanity that is "to be with the other" or "to be together" (Maffesoli, 1987). This perception establishes dialogue with Heidegger's propositions, for whom the community was an essential existential experience. With this, we understand that is from the communitary experience that the subject reaches another spaces, in other words, the world.

In possession of these premises about the community, its evolutions and transformations, next, we will point out different concepts applied to communities.

2.1 Traditional community

The conceptual issue around the definition of the expression "traditional communities or populations" involves the ambiguities found between different existing definitions. However, we can affirm that, throughout the world, the expression "traditional communities" carries an expressive signal that is related to the maintenance of community identity.

When analyzing this meaning, it is observed that there is the definition that traditional populations constitute groups that have conquered or are in the process of conquering the maintenance of their cultural identitary character. This identity related to traditional populations encompasses some aspects, such as the existence of cultural traits that are reaffirmed rethought among the population's social actors, local leaders, community structuring and support in laws to enforce their rights (Almeida and Carneiro da Cunha, 2001).

Another definition of traditional populations was presented by Diegues and Arruda (2001), including indigenous peoples and other human groups that, according to the authors, are characterized by those who have a peculiar form of living according to their experiential regions. Complementing what was said in the previous definition, in this one, it is observed that Diegues and Arruda also consider traditional populations to be those whose human groups carry different cultural traits experienced historically through mutual cooperation and specific relations with the environment, and therefore, these cultural heritage conceived by this population are perpetuated.

To complement these two commented definitions, we also bring a very particular notion of the social sciences. In this context, the concept of traditional populations were absorbed into the legal system when considering both the social and biological diversity (Santilli, 2005).

Making a parallel with Brazil on this especially important issue, the Brazilian government practice, since the 2000's, has been adopting in its public management concepts regarding this type of community and its territory. One movement was the establishment of a an improvement for the sustainable development of these communities. There was also the definition of the expressions "traditional peoples and communities" as the groups that differ culturally and recognize themselves as being distinct from other human groups; and they also make use of their spaces in order to maintain their culture as well as several other aspects existing in their social structure.

In face of these placements, it is perceived two aspects that have been defined regarding traditional peoples and communities, which deserve to be emphasized, which are the territory and access to their natural resources. Therefore, we will now discuss in a little more detail the elements of territory, field of biodiversity, and the symbolic field from the perspective of construction and affirmation of a self-recognition as a traditional human group.

Thinking about the issue of territory, Godelier (1984) reveals to us that it implies in rights related to the access, control, or use of the natural resources by the community management. We belive that territory is the heart where the interactions between these traditional communities and nature occur. Making a allegory, "territory" is a term that can be understood as a "piece of heaven" where nature + space + society constitute a triad that drives people's desire in search of their rights.

If we consider the process of human occupation of territories, we observe that this process gains a very peculiar perception because these community areas are constituted by the human presence that is modifying and constructing the local landscape.

If there is this relation between occupation, permanence, and territory, to think about traditional communities in the path of history is to realize that there has always been a dynamic process of community constitution. Going back to the 19th century, we see that European social theory notifies these so-called traditional territories and their populations in their historical progression, following a path of dilution. This movement pointed out by the 19th century European social theory found its reasons in the development of modernization and, consequently, in the increase in trade that, by going to the search for an urbanization of society, caused a dilution and thus a decline of the community life, which until then was articulated and survived based on its own social collective memory.

In the field of biodiversity, we can think about the composition of biodiversity throught the historical actions that traditional populations or communities have made on the natural environment. Diegues and Arruda (2001) point out that biodiversity is circumscribed in the natural and cultural scopes but that, however, it is culture. Biodiversity is understood as a culture in the sense that it is a knowledge that opens a possibility to traditional populations to understand their biodiverse elements, intellective representation of these elements, as well as handling and interaction with them, as species are being moved and the composition of biodiversity is being altered due to this interaction with humanity. This signifies that the species of flora and fauna were constituted as objects of intellectual apprehension (knowledge), practices (such as domestication and use), and origin for the narrative and belief in the transcendental in traditional populations. Thus, also the elements that make up biodiversity have been serving as objects of exchange, of trade, between human societies.

Therefore, we can say that biodiversity is not only a product of nature. But, yes, in many cases, nature is the product of the intervention of human societies and cultures, represented especially by traditional nonindustrial populations. In this sense, we can also affirm that nature then represents a historical, cultural, and social construction.

Making a clipping within the concept of traditional populations, we can bring to the discussion the expression "indigenous populations". Taking again as an example, particularly Brazil, the use of this expression commonly has the meaning of "ethnicity", referring to peoples who seek to maintain a historical and cultural heritage with an initial milestone prior to the European colonization of the American territory. Following this ancestry and history of territorial occupation, in Brazil this is the condition that led indigenous areas to be demarcated based on their historical right in relation to their territory. With this, there is a more decisive differentiation between the indigenous and nonindigenous populations when we talk about indigenous ethnicities, as well as about the historical and identity maintainance of their societies and cultures.

In the ambit of our discussion about expressions about human populations, it is also worth commenting that there was a deep debate on what the expressions "native populations", "tribal", "indigenous", and "traditional populations" would mean. From the reflections that occurred around the 1990s, a new definition was conceived with The World Bank's

Operational Directive 4.20 of 1991 (World Bank Operational Directive 4.20 on Indigenous Peoples, 2001). This new definition had its meaning expanded, suggesting the change of the term "tribal peoples" to "native peoples", starting to refer to indigenous peoples. Thus, the expression "native peoples" began to be applied to those peoples who live in a specific territory and bearers of several characteristics. These characteristics refer to the following: deep connection with the ancestral territories; self-identification and recognition by other peoples as distinct cultural groups; language, generally different from the official language; presence of their own and traditional social and political illustrations; and production systems especially focused on subsistence (World Bank Operational Directive 4.20 on Indigenous Peoples, 2001).

Even turning to the issue of biodiversity, an essential element that does not fail to guide the discussions when considering the interactions between traditional populations and nature is the relation of these populations with the territory. In this way, we return to the question of territory, which can be understood through the notion of a portion of nature and territory where human societies claim stable rights, as already mentioned earlier in more detail in this chapter. This concept of portion of nature is what strengthens and invigorates the essence of human being. In addition, the portion of nature provides the means of subsistence through the material and immaterial traits produced by each cultural group, dynamizing these relations (Godelier, 1984).

It is interesting to note that the aspects that permeate social relations also integrate what constitutes the territory. That is, the territory depends on social relations and not only on a type of physical environment that is historically used by traditional populations. This signifies that social relations determine social behaviors related to possession, use, and limits of action of each member of traditional populations in relation to the territory they occupy (Cordell, 1982). Then, beyond the territory *per si* there are laws, social norms, which are preserved and respected by the members of a society and which guide the ethics of relations between them.

Faced with this reality in which we understand that the term "territory" encompasses social relations and physical space, another relevant aspect in the definition of traditional societies is how the management of natural resources are done. In traditional cultures, management is usually carried out aiming to preserve natural cycles, being guided by the sustainable use of species of fauna and flora. As a logic of use of natural resources that underlie traditional management practices, we observe the way of using the territory that stands out the existence of knowledge derived from the cultural heritage of the community.

To move forward in our considerations, we propose here a closure of the territory issue considering the field of biodiversity bringing some brief considerations. It is worth reflecting on the concept of territory from the anthropologic conception. The territory then comprises a place where traditional populations are able to express their cultural traces and perpetuate their generations (Santilli, 2005). In this sense, in the conceptualization of territory, it is necessary to realize that there are multiple choices relative to the uses and appropriation of these spaces, propitiating differentiated landscapes, in short, there are particular meanings and feelings about the possession of the territory by traditional communities.

Thus, the territory plays a significant role when we think about the cognitive (Miranda, 2005). By this way, our relations with the culture as a whole and with the elements of nature, invariably, pass through cognitive structures. The importance of place in the mental conception of members of traditional communities occurs once people perceive themselves as individuals and interact in a group coexisting in a specific environment, in which they are inserted and where cognitive social constructions are manifested. These culturally and socially conceived meanings attributed to things in a common territory express individual and community experiences from different populations.

Directing to the field of the symbolic, the territory shelters the cognitive symbols and transcendental imaginary of these traditional goups. Through this symbolism these communities align their interactions with their living environments, with the elements of nature. Therefore, this symbolism strongly implies that traditional communities direct actions regarding the appropriation and use of the territory, guiding the practices, know-how, innovations and production, thus reinforcing the presence of the mythical (Diegues, 1996). In the context of these symbolic and mythical systems, we can try to make their existence more concrete when we approach the elements that populate the imaginary of the traditional communities. By way of example, in Brazil, populating the community imaginary we find magical entities that have already been illustrated by the Brazilian folklorist Luís da Câmara Cascudo (1898–1986) (1972), such as the entity that punish those people who prejudice their rivers and lakes (e.g., "caipora"), those who overfish, beyond the necessary (e.g., "mãed'água", or in English, "mother of water"), and also those people who do not respect and kill in the breeding periods ("tapira").

We presented some forms of expression of this universe of representations, and it would be essential to point out some traditional Brazilian populations that maintain their particular systems of knowledge, practices, and values. Some of these are represented by "quilombola" communities (black groups), besiegers, riverside communities, extractive and indigenous groups. Although we can make this grouping of traditional human society, we must think that it is a reducionist systematization that is imposed on these community conceptions when we turn our gaze to the fact that traditions that are carried by these societies. Despite this aspect of tradition, as the social system can be considered a cultural complex in dynamic movement and in constant transformation, we can observe that in addition to this grouping of traditional societies, there is another grouping, a non-traditional human society, which is incorporated in this process of constant change (Diegues and Arruda, 2001).

It is opportune to bring at this moment the conceptualization of the term "traditional". Commonly the model of traditionality ends up imposing a rigidity to the natural process of social and cultural movement of communities by assuming the principle that the true traditionality is intimately linked to permanence. However, in the community context, the traditional is far from the definition of practices that refers us to cultural traits of the past that have survived to the present day. But, on the contrary, the conceptual understanding of traditional is linked to practices that integrate the movement commented below, where the ethnical changes promote the permanence of the communitary. Thus, understanding the traditional in the complex community sociocultural formation involves the apprehension of the historical movement. In this way, the collectivity of communitary life has a dynamic process, which includes the maintenance, creation, and remodeling of identity traits. The community is constituted based on this cultural tradition, that is, the action of the community collective will find its structure in tradition, including old and new cultural traits.

Then comes another criterion for the conceptualization of traditional cultures or communities. What is important in this process of traditionality is the consciousness that individuals have of being part of a social group. Here, it is evident that beyond the way of life, the essential is the meaning of individual or collective identity. The identity character develops a representation of ethnic group, which is characterized by the relations and perceptions established by the very integration in a system in which it deeps in new perspectives of community identification based on different sociocultural identities, which turn to the consciousness of traditional populations. In this way, the traditional communities through this movement put themselves in the time and space they occupy and go on.

2.2 Neocommunities

In our continuity in relation to the communities, we have notified the movement to revisit the existing traditions from the perspective of new actors who, at the same time they have their own baggage from within the communities, also acquire external experiences that will function as input to construct a new form of community. We point out that there is still a conflicting scenario regarding contact between distinct societies, because they do not always have shared cultural elements, and what has been seen in more recent history is that traditional communities have been valued by industrial societies precisely because they carry some specificities related to their biologic or cultural traits.

We will contemplate some aspects of this type of so-called neotraditional community within a general historical panorama. Reflecting this appreciation, heritage policies and tourism are gaining *momentum* around the world. Commenting in general, since the 1980s the European Union has been working to revitalize the sites that kept in their landscape and cultural traditions traces of the communities that settled there in the past. To this end, public policies have been implemented in the villages, with the aim of rescuing the cultural tradition and for that purpose they have been planning to reinvigorate and remodel the design and the environment of these places. In this process, the community projection of the villages gained a set of actions envolving tourism and also the construction of a policy directed to the protection of the landscape. Therefore, this policy have been consolidating the European identity and we observe that there is an action for individual and also collective well-being and the search for the revaluation of European culture and its immaterial and material heritage.

Another aspect that has grown in relation to the community space has been to revisit and value cultural heritage utilizing the language through orality and other forms of its expression as sources that can reveal the traditions and history of these people and their sites. This appreciation of history occurred especially due to the multiplicity of narratives and searches on histories at a local scale. According to L'Estoile (2001) and Sagnes (2002), these local historiographic narratives made about various localities, by local people (scholars or not), have gone into encounter to the tourism desire. Thus, there is the construction of narratives of the facts or versions of them related to the local histories. It is interesting to note that these local historical narratives made by inhabitants of the villages themselves were in line with the process of revaluing the traditions of the communities.

After constructing this path of reasoning, let us look at this issue of the revalorization of the community traditions from another perspective. In the case of Europe, mainly in the southern regions of this continent, this movement the necessity to scape from the economic crisis that the countries of the world were going through (Figueiredo, 2003). In this sense, the revalorization of European rural traditions would not have focused solely on respect for the "other", the villager's practice and way of life. It was an economic plan to increase revenues, in which the rural producer was stimulated to utilize the spaces considered unproductive, thus complying with this economic plan and, at the same time, also working in favor of their traditions aiming at the objective of cultural tourism (Butler and Hall, 1998).

This development policy can find its roots in the cultural phase of the capitalism and this way to see the *phenomenon* of rediscovery of traditions in communities may be observed more comprehensively. From this point of view, the aforementioned new economic policy promoted a revalorization of traditions by giving a monetary value from the cultural traits and feelings experienced in the community plot (Terranova, 2000). Thus, the economic policy implemented led to the revalorization of traditional communities alongside the appreciation of the currency, thus moving the monetary scenario in these spaces.

In addition to the economic need, the creation of a new monetary value, the revalorization of communities also comprised the existential plan (Bauman, 2003). In this aspect, people started to aspire to one of those that would have been the landmark of modernity, that is, the determination to live in large cities embracing the expectation of having freedom through a way of life unknown to others, that is, a private life. In this new condition of way of life, people exchanged the security of life in a small community for freedom in urban centers. Nevertheless, over time, due to an existential emptiness, people started to rethink this way of life in large cities and seek a return to greater communion in social relations, in addition to other aspects experienced in the community plot such as security and their financial support. Especially security has transformed in one of the most difficult issues faced by those who live in a large city. So, what is observed is that, as a backdrop of the process of revalorization of the traditional community, there was the desire to have again security conferred by the closest social relations.

Several motivations have boosted the revalorization of traditional communities, and in this scenario, we can see the contemporary *phenomena* of cultural reconstruction that involves two dimensions: that of communities with their traditions and knowledge and another that of external agents that seek to rescue cultural traditions, incorporating a new meaning to these cultures.

Thus, what we perceive is that neocommunities reorganize and reupdate the social-cultural context by considering two dimensions, that are, past-present and internal-external traces of the social memory. This configures that, in a same community space, there is an action carried out by external agents who implement new practices to traditional knowledge, as it was said, and this new manifestation of life becomes part of the community as a whole.

In this way, the conception of neotraditional communities involves the understanding that these populations have, at the same time, traditional knowledge, an acquisition of new traditional knowledge, as well as new external knowledge. These nuances of typologies of knowledge that characterize neocommunities can give them an ability to be more resilient to change (Begossi, 1998; Berkes, 1999). Nevertheless, although neotraditional knowledge has this configuration, as we understand traditional knowledge as essencially dynamic, in constant movement, we also understand that traditional communities in contact with other groups may, consciously or not, vary in the degree of acceptance regarding these external and new knowledge (Medeiros and Albuquerque, 2012).

Self-recognition by parts of human populations as being neotraditional is a process that sometimes involves the discovery of an already existing identity and, also, its updating, through the self-perception of their cultural group as having a social memory and a unique rhythm of lives. We can also say that it stems from interactions established between the neocommunities and the urban-industrial society and that, over time, these relations have been generating divergences due to different conceptions of structuring social life.

It is also relevant to think that, with the appearance of neocommunities, the existence of a space for intercultural interaction will be updated in which traditional and external agents will exercise interculturality. Nevertheless, when we consider the perspective in which we distinguish the use of the term "neocommunities", we can find some problematic situation. This is due to the fact that the relations between different agents in a unique interaction space commonly does not happen in a complementary way, in a conscientious form. In view of this, the relation between traditional and external agents ends up generating conflicts of interest. In addition, this conflicting situation of the relations established between the different agents occurs because in many communities what is taken into account is the possibility of building a new identity that is not always welcomed by each member of the two societies (internal and external members of the so-called neocommunity).

Discussing about identity, we can bring to our reflections the ideas of the German sociologist Maximilian Weber (1864–1920) (1997) that can contribute to the understanding of the *phenomenon* of neocommunities. Weber went further when his contribution emphasized the subjectivity of the individual in a community, going deeper into the subjective feeling that each person brings, even living in community. From this placement, Weber believed that individuals could form a community. Thus, the community arises from an internal movement of the subjects, who in their locality self-recognize their community identity.

Amog the concepts of communities we can propose a parallel between this subjectivity of Weber's understanding with that of Tönnies's, which we have previously commented. Although Weber does not devalue aspects of social relations (ancestry and occupation of the same territory), he highlights the people's subjective feeling of belonging to a group. This is what confers a specificity to Weber's consideration about community. In this sense, we can also mention that Weber defines ethnic communities, such as those that comprise a subjective belief in a common origin, based on memories and life practices, regardless of the cosanguinity (Weber, 1997).

The question about Weber's conception is that there is a symbolic aspect that will establish the ethnic community and according to this thought, some researchers considered it as reference to their studies of community. Among these studies, we can mention those undertaken by the Norwegian anthropologist Friedrik Barth (1928–2016). Barth's contributions will include discussions on ethnicity policies undertaken about traditional communities, especially in the country of Brazil.

The focus given by Barth to ethnic communities shows that these populations have other elements to be analyzed beyond territoriality, bonds of consanguinity, and ancestry. Barth postulates in his book *Los Grupos Etnicos y sus Fronteras - La Organización de las Diferencias Culturales* (Barth, 1976) that ethnic groups establish their identity in the interrelation with other groups considering the symbolic dimension. In this way, the ethnic community would be structured in subjective elements, which is opposed to diacritic or objective traits (for example, language, customs or kinship). Thus, for Barth, ethnic groups build their identities through the activation of those objective traits they elect to be significant for each context of social interaction.

It then occurs that Barth considers that communities confers to themselves identity signs to the extent that there is a differentiated relation with other social groups. Thus, ethnic identity has as "dependent variables" the nature and extension of the relations between distinct social groups, which deals to a less dependence on intrinsic cultural differences for the constitution of ethnic identity. In contrast to Tönnies' proposal, for Barth the definition of identities is not centered on issues of community of place or community of blood, but the focus is on identifying the context in which they occur, that means, analyzing the nature and extention of these social relations.

After these comments that inform us well about the conception of neotraditional communities, we will continue with a brief analysis on the insertion of this community life in the American continent.

Initially, focusing on Brazil, during the 1960s there was a change in the relation between the countryside and the city. With this process, society changes its predominance, being integrated in major by an industrial and urban society. As a result of this modernization, there was a progressive disregard in respect to traditional communities by the fields of research and public and governmental action. With the entry in the 1990s, however, there was gradually a return to what is the local community. This movement occurred through the renewed interest of academic researches, as well as the action of public policies that were implemented at that time.

In the same period, in the European context, local communities also returned to the field of discussion from the inversion of demographic trends caused by the revalorization of the rural and local communities. This movement in Europe was largely due to tourism at a local level, as well as to the heritage policy adopted, as we mentioned earlier.

In Latin America, the return to the community found different reliefs. The scenario in the countries of this region, for the most part, is characterized by the so-called recognition policies of ethnically distinct groups. These policies occurred with a process of precariousness and forgetfulness of identity signs, including ceremonies, rituals, native languages, body paintings, healing practices, among many other particular cultural traits of local communities. This movement of abandonment of these identities were interpreted as a sensitive indication of this process of cultural changes what lead to the disappearance of cultural groups from several Latin America countries (Saez, 2003). Nowadays, however, these identity signs of cultural groups, even those considered over, have been reborn and gain strength as a mark of resistance in these communities. This *phenomenon* has been called by different terms, such as rescue, revitalization, in short, everything that brings us to the remodeling of the traditions of local communities. In general, this represents initiatives to rescue knowledge and practices that were abandoned during this process of multiculturalist drift, which would be configured as the invisibility process of local communities and which will now be rethought (Saez, 2003).

There is yet another fact that could be considered in the analysis of the rediscovery of traditional communities. It is characterized by the involvement of people with the culture issues through organizations called nongovernmental organizations (NGOs), which work especially within the community in social and cultural projects, creating funds in favor of the development of the community, and also in the articulation with social movements. From this action of the NGOs, community culture began to be incorporated within an institutional system and began to be considered.

In this process of rescuing we find other many types of situations, which, even though they are different from each other, signal the social importance of indigenous and black populations. Other points in common among these various situations involved in this rescue are the revitalizations of traditions revealed through identity signs or marks. As an example of this identity rescue movement, we can mention the revitalization of identity signs through body paintings among indigenous groups in Brazil, a fact that we were able to observe especially among the indigenous people from the coast of Paraíba (Northeastern Brazil), when we taught ethnobiology for the training of indigenous biology teachers.

Still having Brazil as an example, the State established official norms contemplating the institution of ethnic communities, such as the "quilombola" identity. These so-called remaining "quilombo" communities will rise from the law, more precisely, from the Brazilian Constitution of 1988. Article 68 of the constitution postulated that to the so-called remnants of "quilombo" communities, "which occupied their territories, would have the definitive recognition of property" (BRASIL, 1988). It also added, stating that in such cases, the State should issue titles to these remnants of communities, whose definition reached a broader meaning encompassing those who recognize themselves as "quilombola", being part of the "quilombola" community. By this constitutional force, in Brazil, the "quilombola" communities could then obtain their land ownership.

However, this policy did not only address the land dimension. In addition to this aspect, throughout the governments, there was a policy aimed at "quilombolas" that was opened to the dissemination of cultural manifestations from black populations. Another movement that also found space was the recognition of the need to maintain the material and immaterial heritage of the "quilombolas". In view of all this historical unfolding, taking as an example the "quilombolas", we can say that through politics and social representations through its movements, there was a revalorization of the values conferred on these ethnicities. This openness in thinking about these elements that form human groupings allowed new connections and articulations between social memories, identities signs, and external agents to be realized.

Expanding our perspective of observation, we can say that in the neo-communities the so-called "external agents" influence directly on changes. However, these agents are not characterized as those who bring advanced elements to the communities. On the contrary, with their actions to rebuild ethnic communities, traditions are maintained.

Thus, in this scenario, it is important to say that neocommunities form spaces in which external and traditional agents have the possibility to interact and exchange experiences. Therefore, interculturality occurs in these territories that become a field of possibilities of interaction between agents from different origins that lead us to see the neocommunities as being places that privilege those cultural exchanges.

3. Brief closing of ideas

What guided the discussions in this chapter was the view that the identity of a given ethnic group could not be understood by taking only objective elements, but to perceive that subjectivism has a preponderant place, including in this regard, knowledge, cultural values, the transcendental dimension, among other aspects related to the identity of the social network.

Thus, ethnic groups have the autonomy to define their identity based on their own relations and identification of diferentiated traits when a parallel with other cultural groups is made, which implies the absence of pre-established patterns and the abscence of general denomination of what would be defined as ethnicity. That is, ethnicity is not framed in any preconceived scheme, since its conception is guided by people's self-attribution, which confers them freedom to choose what gives them a sense of belonging.

What we would like to leave as the final idea of this chapter is that the marks of ethnicity are in constant movement, are changeable, and by this nuance, they are placed in permanent questioning by the ethnic group itself. This dynamic is what permeates the existence of ethnic groups throughout history, and, thus the criteria of ethnicity accompany this tendency of freedom and autonomy for the eligibility of a community identity.

References

Albuquerque, U.P., 2005. Introdução à etnobotânica, 2. ed. Editora Interciência, Rio de Janeiro.

Almeida, M.W.B., Carneiro da Cunha, M., 2001. Populações tradicionais e conservação ambiental. In: Capobianco, J. P. et al. (Orgs.), Biodiversidade Na Amazonia Brasileira. Estação Liberdade/Instituto Socioambiental, São Paulo.

Alves, A., Albuquerque, U.P., 2005. Exorcizando termos em etnobiologia e etnoecologia. In: Alves, A.G.C., Lucena, R.F.P., Albuquerque, U.P., Orgs (Eds.), Atualidades em Etnobiologia e Etnoecologia 2. NUPEEA, Recife, pp. 11–25.

Ander-Egg, E., 1978. Introducción a lãs técnicas de investigación social: para trabajadores sociales, vol. 7. Humanitas, Buenos Aires.

Babini, J., 1957. El saber. Nueva Visíon, Buenos Aires. BANCO MUNDIAL. Diretiva Operacional 4.20, de setembro de 1991.

Barth, F., 1976. Los grupos etnicos y sus fronteras. La organización de las diferencias culturales. Fondo de Cultura Económica, México.

Bauman, Z., 2003. Comunidade. In: Zahar, J. (Ed.), (Rio de Janeiro).

Begossi, A., 1998. Resilience and neo-traditional populations: the caiçaras (Atlantic Forest) and caboclos (Amazon, Brazil). In: Berkes, F., Folke, C. (Eds.), Linking Social and Ecological Systems: Management Practices and Social Mechanisms for Building Resilience. Cambridge University Press, Cambridge, pp. 129–157.

Berkes, F., 1999. Sacred Ecology. Taylor & Francis, Philadelphia.

Boyd, R., Richerson, P.J., 2005. The Origins and Evolution of Human Culture. Oxford University Press, Oxford.

BRASIL, 1988. Constituição da República Federativa do Brasil de. In: Presidência da República, Brasília. BRASIL. Lei Imperial número 3.353, sancionada em 13 de maio de 1888. Paço do Senado, Rio de Janeiro.

Burke, P., 1989. Cultura Popular na Idade Moderna. Companhia das Letras, São Paulo.

Butler, R., Hall, C., 1998. Tourism and Recreation in Rural Areas. John Willey & Sons, Chichester.

Cascudo, L.C., 1972. Dicionário do folclore brasileiro. Editora Ouro, São Paulo.

Cavalli-Sforza, L.L., Feldman, M., 1981. Cultural Transmission and Evolution: A Quantitative Approach. Princeton University Press, Princeton.

Cordell, J., 1982. Locally Managed Sea Territories in Brazilian Coastal Fishing. FAO, Roma.

Campos, M.D., 2006. A cosmologia dos Caiapós. Sci. Am. Bras. 14, 62–71.

Diegues, A.C., 1996. O mito moderno da natureza intocada. Hucitec, São Paulo.

Diegues, A.C., Arruda, R.S., 2001. Saberes tradicionais e biodiversidade no Brasil. USP, São Paulo.

Escobar, A., 2005. Más allá del Tercer Mundo. Globalización y diferencia. Icanh-Universidad del Cauca, Bogotá.

Figueiredo, E., 2003. Um rural para viver, outro para visitar: o ambiente nas estratégias de desenvolvimento para as áreas rurais. Tese (Doutorado em Ciências Aplicadas ao Ambiente). Universidade de Arveiro, Arveiro.

Godelier, M., 1984. L'idéel et le matériel. Fayard, Paris.

Harris, M., 1964. The Nature of Cultural Things. Random House, New York.

Harris, M., 1969. Emics, Etics, and the New Ethnography, the Rise of Anthropological Theory. Routledge & Kegan Paul, London.

Harris, M., 1976. History and significance of the emic/etic distinction. Annu. Rev. Anthropol. 5, 329–350.

Harris, M., 1990. Emics and etics revisited. In: Headland, T., Pike, K., Harris, M. (Eds.), Emics/Etics: The Insider/Outsider Debate. Sage, London, pp. 48–61.

Hewlett, B.S., Cavalli-Sforza, L.L., 1986. Cultural transmission among Aka Pygmies. Am. Anthropol. 88, 922–934.

Heyes, C.M., 1994. Social learning in animals: categories and mechanisms. Biol. Rev. 69, 207–231.

Hoppit, W., Laland, K.N., 2008. Social process influencing learning in animals: a review of the evidence. Adv. Stud. Behav. 38, 105–165.

Hussey, S.C., 1989. An Annotated Bibliography of Publications Using the Emic/etic Concept, a Text File in a Database Format (DBF) on Diskette for IBM and Compatibles. Summer Institute of Linguistics.

L'Estoile, B., 2001. Le goût du passé. Érudition locale et appropriation du territoire. Terrain 97, 123–138.

Latour, B., 1999. Pandora's Hope. Essays on the Reality of Science Studies. Harvard University Press, Cambridge.

Lifschitz, J.A., 2011. Comunidades tradicionais e neocomunidades. Contra Capa, Rio de Janeiro.

Maffesoli, M., 1987. O tempo das tribos. Forense Universitária, Rio de Janeiro.

Medeiros, M.F.T., Albuquerque, U.P., 2012. Dicionário Brasileiro de Etnobiologia e Etnoecologia. SBEE/NUPEEA, Recife.

Mesoudi, A., 2007. A Darwinian theory of cultural evolution can promote an evolutionary synthesis for the social sciences. Biol. Theory 2, 263–275.

Mesoudi, A., Whiten, A., Ialand, K., 2004. Perspective: is human cultural evolution Darwinian? Evidence reviewed from the perspective of the Origin of Species. Evolution 58, 1–11.

Mesoudi, A., Whiten, A., Laland, K., 2006. Towards a unified science of cultural evolution. Behav. Brain Sci. 29, 329–383.

Mesoudi, A., Whitten, A., 2008. The multiple roles of cultural transmission experiments in understanding human cultural evolution. Phil. Trans. 363, 3489–3501.

Miranda, M.L.C., 2005. Organização e representação do conhecimento: fundamentos teórico-metodológicos na busca e recuperação da informação em ambientes virtuais. In: Tese (Doutorado em Ciências da Informação) – Escola de Comunicação. Universidade Federal do Rio de Janeiro, Rio de Janeiro.

Nazarea, V.D., 1999. Ethnoecology: Situated Knowledge/Located Lives. University of Arizona Press, Tucson.

Pike, K.L., 1954. Language in Relation to a Unified Theory of the Structure of Human Behavior. Summer Institute of Linguistics, Glendale.

Pike, K.L., 1978. Social interaction as the break-in point for the analysis of verbal behavior. In: Proceedings of the Twelfth International Congress of Linguists, Viena, pp. 739–741.

Pike, K.L., 1990. On the emics and etics of Pike and Harris. In: Headland, T., Pike, K., Harris, M. (Eds.), Emics and Etics: The Insider/Outsider Debate. Sage, London.

Posey, D.A., 1999. Ethnobiology and ethnoecology in the context of national laws and international agreements affecting indigenous and local knowledge and traditional resources – with some comments on Latin America. Etnoecologia 6.

Richerson, P.J., Boyd, R., 2005. Not by Genes Alone: How Culture Transformed Human Evolution. The University of Chicago Press, Chicago.

Rosa, M., Orey, D.C., 2012. O campo de pesquisa em etnomodelagem: as abordagens êmica, ética e dialética. Educação e Pesquisa 38 (4), 865–879.

Ruiz, J.A., 1979. Metodologia cinetífica: guia para eficiência nos estudos. São Paulo, Atlas.

Saez, O., 2003. Um balance provisional del multiculturalismo brasileño. Los Indios de las Tierras Bajas em El siglo XXI. Antropologia em Primeira Mão, PPGAS, Universidade Federal de Santa Catarina.

Sagnes, S., 2002. Le passe dês historiens locaux. Ethnologies comparées (Revue électronique) 4, 1–24. http://alor.univ-mon-tp3.fr/cerce/revue.htm.

Santilli, J., 2005. Socioambientalismo e novos direitos: proteção jurídica à diversidade biológica e cultural. Peirópolis, São Paulo.

Terranova, T., 2000. Free labor: producing culture for the digital economy. Soc. Text 18 (2), 33–58.
Tönnies, F., 1947. Comunidade e sociedade. Editora Losada, Buenos Aires 1881.
Tönnies, F., 1995. Comunidade e Sociedade. In: Miranda, O., Org (Eds.), Para Ler Ferdinand Tönnies. Edusp, São Paulo, pp. 231–352.
Trujillo Ferrari, A., 1974. Metodologia da ciência, 3. ed. Kennedy, Rio de Janeiro.
Vandebroek, I., Reyes-Garcia, V., Albuquerque, U.P., Bussmann, R., Pieroni, A., 2011. Local knowledge: who cares? J. Ethnobiol. Ethnomed. 7, 35.
Viertler, R.B., 2002. Métodos antropológicos como ferramenta para estudos em etnobiologia e etnoecologia. In: Amorozo, M.C., Ming, L.C., Silva, S.M.P. (Eds.), Seminário de Etnobiologia e Etnoecologia do Sudeste, 2001. Anais. UNESP, Rio Claro, pp. 11–29.
Warbunton, H., Martin, A.M., 1999. Local People's Knowledge in Natural Resources Research. Best Practice Guideline. Socio-Economic Methodologies Programme. DFID, London.
Weber, M., 1997. Economia e sociedade. Fundamentos de la sociologia compreensiva. Fondo de Cultura Económica, México.
World Bank Operational Directive 4.20 on Indigenous Peoples, 2001. https://www.ifc.org/wps/wcm/connect/4da94701-07cc-4df3-9798-947704a738d4/OD420_Indigenous Peoples.p df?MOD=AJPERES&CVID=jqewORB. (Accessed 26 October 2020).

CHAPTER EIGHT

Teaching historical ethnobiology

Contents

1. Introduction to the study of historical ethnobiology

The study of ethnobiology in school or higher university education is not even present in disciplines and/or courses of areas such as biology, sociology, or anthropology, for example.

By conjugating ethnobiology with history, we enter into the universe of teaching history from the perspective of historical ethnobiology. With regard to the study of history in school and higher education, this study was regularly conditioned to a pattern in which the teaching process was centered in politic events. This type of education utilized memorization as a strategy of learning about people (their names), time (dates), space (locations), and some specific terms that signaled the nature of studied facts. This teaching-learning model reflected what is denominated of traditional

Historical Ethnobiology
ISBN: 978-0-12-816245-3
https://doi.org/10.1016/B978-0-12-816245-3.00008-1

teaching, which represented the conjugation of two approaches: one in the field of historiography and the other in the field of pedagogy.

When we think about the study of traditional historiography at different levels and not only focused on ethnobiology, we can say that it is based on a research model developed in the second half of the 19th century. The emphasis given by traditional historiography is focused on understanding the role of the State in the learning about the memory of the past. Thus, foll-lowing this traditional education, official documents were used as sources for the study of history, which was understood as a discipline to promove the understanding about the constitution of the State. Turning our gaze to the practice of this type of historiographical study, there is a model of historical analysis that results in a description of social actors of great prominence specially in the public scenario, such as military, revolution, and political leaders. Sometimes, traditional historiography also addresses in its teaching practice the achievements of social actors such as scientists, religious, artists, among others, but always has as a major plan of analysis of a historical period regarding the political context.

From the perspective of pedagogy, its traditional current has students as non-active subjects in the teaching-learning process. The focus given by traditional pedagogy is on memorizing the content presented by the teacher to the students in the classroom. Thus, there is no learning in which the focus is on understanding the contents studied. This approach characterizes the school space as a place where students' individuality cannot be expressed. This process ends up causing the massification of students, because space is lost as well as the exercise of individual freedom to develop their imagination, their own thoughts, experiences and conclusions about the themes focused.

Contrary to this traditionality in teaching, when analyzing current textbooks in the areas of history, biology, pedagogy, and also ethnobiology, we observe that there is a demand for not adopting the premises of the traditional current anymore. In this way, we can observe that teaching practice has surpassed traditional historiography through the insertion of new objects of study in formal education, such as customs, traditions and daily life. Another path that has also been undertaken has been to consider old objects under a new meaning. In this aspect, historiography may well intersect with ethnobiology when considering in its teaching, for example, a perception of traditional and indigenous peoples not in an exotic way, but rather as people who carry different worldviews, wisdoms, and cultural traditions. On the pedagogy side, the surpass of the traditional current has been taking place with the adoption of proposals that use both memorization and discussion to establish a reflection on the contents that are being focused on.

We will try to address these issues in a perspective in which we will relate the teaching of historical ethnobiology to current historiographic trends. We will also present a roadmap with suggestions for activities that can be adopted and adjusted to the reality of each teaching situation, thus being able to assist in the planning of didactic sequences by the teacher.

2. Studying historical ethnobiology

The great objective of historical ethnobiology is to encourage the student to seek and formulate a better understanding of the society and its relation with nature. As social and cultural beings, this understanding becomes essential to people. And this search for understanding encompasses individual identity (anxieties, memories, affections, knowledge, etc.), collectives, and various identity scales such as local and national. All these levels of identity constitute history.

As presented in a previous chapter, we can say that history is dedicated to the study of human being, looking for the vestiges of social memory that are related to the social plot, in its different temporal and spatial dimensions. Its intention is to reflect on various aspects of human life over time, seeking to identify changes and permanence addressing cultural traditions, scientific institutions, politics, economy, relation with the environment, and so on. Historiographic thinking not only turns to the human being as an isolated being, but also considers the category of temporal duration of the spheres that involve human life.

Many times, we create the expectation that history will answer all inquiries that we have in relation to fulfill certain actions. However, we must consider that the ideas and practices made in the remote or more recent past have a repercussion in tune with the broader historical context, typical of the moment when the developments resulting from this knowledge and practical action occurred. Therefore, it is not possible to transpose these vestiges of memory in an unthinking way to our time, since our contexts are different from that specific moment in history. Thus, the necessary reflection for the ethnobiologist is that a practice may have produced developments in the past and will not necessarily follow the same pattern if it is reproduced today. In this way, the investigation of the past operates to allow us to reach a broader understanding of the complex of knowledge and practices of people who, organized in society, give their experiences meaning and feelings.

Despite this distinction between societies by the mark of time, it is also a fact that, without history, we would be without a time frame and would not know about our vestiges of social memory, which gives our identity character. This is due to the fact that the experienced in the past time to some degree are reflected in the passage of time because they are intimately linked to the essence of what comprises the human.Aspects such as perception of nature, investigative spirit, and the joy of a discovery are common in all times of history, although under various forms of expression in each of them. With the knowledge of past experiences, with the search for their understanding, a path opens that allows us to reach a more in-depth reflection on these different aspects of life so that we can make decisions that allow us new ways of living nowadays.

To this end, the ethnobiologist dedicated to the study of the past will gather documentary elements (historical sources) for the critical analysis of the information contained in them and, through this procedure, historical memory is reached. Thus, what we intend to affirm is that the research must be conducted through methodological procedures that enable an in-depth analysis of the documents that revolve around a specific topic being studied, thus seeking to verify whether the information is valid or not and whether the raised postulations are justified. Thus, what we intend to state is that the research should be conducted with procedures. A caveat is made by historian Allan Megill (2007) who stresses that is dangerous when history collects its clues simply from preserved individual memories or from the idea of its unfolding as a narrative of commemoration of the facts. Megill continues to tell us that we should also not see history as a mode of tradition, since the submersion of history in memory, in the commemoration of facts, in tradition tends not to exercise its critical action.

Thus, starting from an investigation of the documents of the time when one wishes to situate the field of study, we must always keep in mind that in the process of remembering something that occurred, of working with memory, there is a difference between what we remember and what occured in reality. For this reason, the analysis and the construction of a historical memory should be guided by a deep investigation of the facts that occurred, indicating historical changes and permanences, relating events, institutions, and processes of the past to their respective historical contexts.

3. Brief historiography of historical research

Considering the study of history, nowadays we can find a variety of theoretical perspectives. However, for the understanding of how research is done nowadays, it is necessary to know some of the main moments that integrated the process of formation of historical knowledge as a field of knowledge. Punctuating this history is essential so the ethnobiologist does not lose the integration between teaching and research considering the perspective of the development of historical ethnobiology.

To the extent that, for a person, memories related to their own experiences are fundamental for their identity formation, in the same way, but on the scale of collective memory, the various human groupings from different regions also seek to build narratives about their past, which will form the traces of social memory. This individual and collective movement in search of identity has always occurred because people seek in the facts of the past to find and understand their identity, their belonging and, thus, the recognition of their cultural traits.

Following this need to build an identity, some peoples produced documents throughout their existence with a historical awareness similar to the current investigative exercise of historians. However, despite this common interest among people for their community past, we must consider that the production of historical knowledge, through the construction of historiographical narratives, is not available in all communities. Unlike the constitution of an academic historiography, some peoples have a narrative of the past that is based on myths, on the transcendental dimension. This mythical consciousness present, for example, between Greeks and Romans sometimes was not the only form of constitution of the historiography of that civilization.

In this context, we can mention the Greek historian of the antiquity named Herodotus and another historian known as Thucydides. The first one developed her craft based on an eye for detail and also on the systematic of verifying the validity of the information (Herodotus, 2014). These two constitutive elements of Herodotus' practice continue to be fundamental to the practice of historiographical research until today. He was concerned with a very comprehensive history, in making a general description of what he saw and heard. This *modus operandi* allows us to make a search for an acuity in information and a separation between what is the truth (in fact occured) from what is false (stories, talles). But Thucydides, unlike Herodotus, sought in his narratives to explain and interpret the facts *per se*, going beyond their descriptions (Tucídides, 2001). Thus, he sought to make a detailed analysis of the information he had about a determined fact. This conduct is also configured as an investigative practice that is valid for today.

Therefore, the ethnobiologist who wishes to make the historical cut in his research and wants to work with the teaching of historical ethnobiology should not only register the fact, but also, should be concerned with analyzing it from a perspective that considers its original context, thus leading to its understanding.

Thus, considering the historical process of the formation of historiographical thoughts and practices, we can affirm that since antiquity we have encountered historians. These thinkers wrote about history, however, without presenting theories about its nature and meaning. Because of this aspect, the narratives elaborated by these historians of antiquity are normally considered without depth.

A principle of thought about the theory of history can be situated in the theoretical postulations made by the thinker Aurélio Augustine. Living in a period of transition between the ancient and medieval world, Augustine was a Christian thinker who contributed with theoretical postulations on the historical process and the essence of historiographical knowledge production. Augustine's focus was to understand history based on the Christian faith (Agostinho, 1984). It should be noted that Christianity contributed to our understanding of the history of mankind by structuring it based on two facts. This is because Christianity has presented a general history in a continuous form, starting from an initial fact that is the creation of world, and the end, which comprises the coming of the kingdom of God. Added to these two fundamental characteristics, there is an essential moment for the historiographical understanding of humanity, that is, the birth of Christ.

Although Augustine played a determining role about the thought of the theory of history, there was still no concern for the history in a broader sense, that is, what he did was a quest to understand history in order to better understand the Christianity. Only in the 18th century a systematic production of knowledge emerge, aimed at establishing postulations related to the meaning and nature of history. Among the 18th century thinkers who devoted themselves to this issue, we can quote Voltaire, for whom history should be understood as another narrative, different from fable (the habit of telling stories) (Voltaire, 1769). Thus, for Voltaire, the procedure of bringing together the factual and the imaginary (fantasy) in the same discourse should be a practice to be abandoned so that historical knowledge could gain historiographic independence.

Through these placements, Voltaire signaled his concern with the veracity of facts, that means, of historical facts. However, the model of history he proposed had its conception focused on the objectification of history as a vehicle that could demonstrate the moral conduct for human societies.

Advancing in history, already in the 19th century, narratives began to be carried out in which the history started to have a scientific character. The search for this autonomy in the production of historiographical knowledge led historians to proceed with the already mentioned distinction between the real fact and fiction and, in addition, they started to disconnect it from any moral conception that could restrict it. In this sense, the historian Leopold Von Ranke made an essential contribution by postulating that history must dedicate itself to presenting the facts of the past adhering to what happened and, for that, it must be based on a classification of importance among the typologies of sources Von Ranke, 2011. Among the documents available to historians, Ranke gave written documentation greater reliability than other sources, such as oral reports. In this way, human groupings from the past that did not know the writing should be placed as belonging to prehistory. And history would then encompass those who were endowed with this ability and who would use various supports as basis for conveying the most varied facts through writing.

Another important aspect of Ranke's propositions was the establishment of external and internal criticism of the sources. External criticism would be developed in order to achieve the document's authenticity, and internal criticism would be aimed at checking gaps or contradictions in the content. If any of these conditions were identified in the document under analysis, historians would need to identify the reality not presented in the explicit content present at the source.

In the 19th century, the conception of history and historical research had a significant advance, creating the conditions for this area of knowledge to detach itself from the condition and association with a literary genre and establish itself as a science. It is also necessary to comment that the model proposed by Ranke established a methodological procedure aimed at objectivity in historical research. Despite this, the model proposed by Ranke presented two problems.

One of the problems would be that for the accomplishment of the historian's work, it would be necessary to establish a limit to define what history could or could not study. The necessary cutout for historical research becomes fundamental to not confer equal weight of importance for every detail of the past. This procedure allows a comprehensive synthesis of the theme that is researched and this restriction would be limited to the study of politics and religion. In this sense, for example, the alteration of the landscape of a rural property would not constitute a historical fact because it does not have a very broad social repercussion. But the alteration of nature in a region representative of a natural biome by collective human actions would be a historical fact, as it would have political relevance and, thus would interest the whole society.

The other problem was the inability to understand peoples of the past who did not organize their societies into states. The 19th century was marked by being the time of consolidation of the model of modern state and, at that time, historiographic studies were limited to peoples who had some similarity with this type of social organization. Following this understanding of what to study or not, researchers dedicated themselves to studying ancient peoples who were sedentary, dominated writing and organized themselves politically into empires, thus leaving out of scope those peoples who, although significant, did not have these characteristics.

Nowadays, there is an appreciation of the theoretical contributions of 19th-century historiography, called by many of traditional or positivist historiography. Ranke's model of history is recognized for his contributions, but contemporary historiography has developed a critical analysis and a perception of the gaps in this tendency and other proposals. Through this perception of the traditional model, several new theoretical proposals emerged in the 20th century.

Despite having contributed decisively to the improvement of methodological standards in historiography, positivist history favored politics and official documents. With this positivist attitude, many other traits of memory ended up being excluded by and from history, and there was no opening for different approaches and narratives.

In 1930, an important counterpoint to these characteristics of making a positivist history arose in France. That year, there was the creation of a French magazine, which was called *Annales d'Histoire Économique et Sociale*. In the following decade, in 1940, this magazine modified its title and was called *Annales*: *Economies*, *Societés*, *Civilizations*. The movement of this publication resulted in the creation of a current of historical studies known by *École des Annales*. Throughout the development of this school, historians with different positions have been passing through it and leaving their contribution, converging their ideas to some points in common. These shared aspects would be the search for new typologies of sources and objects of analysis and, also, new ways of constructing the historical narrative. In addition to French historiography, another significant contribution was given by English social history, of Marxist inspiration, as well as other theoretical currents. The repercussions of these schools, *Annales and English Social History*, were intense to the point of generating a new way of seeing historical studies in almost all the world. One of the significant changes was that any fact no longer needed to be linked to politics to be historiographically important and thus considered. This change in conception was feasible due to the concept of structure. For this concept, any trace of memory would be considered to be of importance to history because it is a trace of memory, a representative part of a structured totality.

Already in the 1960 and 1970s, and for a long time, having the writing of a new history, with new outlines was related to the quantitative approach, with the use of statistical methods, to perceive changes and permanences in certain time intervals. This same idea of innovation in the historical narrative also appeared in treatments aimed at the social economic structure. It was intended, therefore, to achieve what was essentially important in the past and, thus, to know about which was the structure that made it possible to make sense of historical events.

At the end of the 1970s, however, this scenario was altered by the realization that quantitative history had conclusive results that did not reflect all the effort undertaken in the research, and it is noted that there was a distancing from the reader when complex results of statistical and economic calculations were presented.

From that moment on and in recent decades, another strand of historical studies has developed, known as cultural history. In this perspective, there is an attention to the daily history of people who had their existence in the past. Or, microhistory works have also been undertaken, being devoted to more reduced objects, such as a community in a specific place or a naturalist. Unlike paying attention and dealing with large regions and for a long period of time, and to illustrate we cite as an example Brazil, a vast territory that had a long colonial period, preference has been given to the study of small communities and to a more closed time frame. With this movement, researches have been able to gain in-depth analysis when dealing with less extensive objects of research. In this same line, themes to be considered were also inserted. More varied aspects have become possibilities for historiographical study, such as colonial processes around a particular species in a specific locality, actions of a social actor of African matrix in relation to the lord and the use of plants to curse him. Furthermore, in a context beyond ethnobiology, we could cite popular art, popular social actors, and popular religiosity as examples of this extension of the theme for historical investigation.

Two strands were thus established: the researchers who are dedicated to the concept of structure and those who emphasize culture and daily life. The former give their textual production an analytical aspect and the latter choose to compose their historiographical discourse in the form of a narrative.

Contemporaneously, there is a tendency to work with objects whose delimitation seeks to restrict the amplitude of the theme, and whose emphasis is placed on cultural practices. The practice of making microhistories or histories of everyday life does not imply having a diffuse interest in any object simply because it integrates social memory. Taking as a study plan a given community, for instance, is of interest to history in the measure that this object of analysis finds its meaning through its articulation with a structure. The studies of local scale contribute in this way, to propose links between historical information, allowing for confirmation or rectification of this information that emerged from the explanatory models. Thus, it is not up to local studies to fail to consider explanatory models, because disregarding the notion of structure is equivalent to giving up the basis from which historical knowledge is thought and constructed.

Taking as depart the commented models, such as traditional history, with emphasis on the history of events and perspective of political leaders, structural history, which emphasizes long-term studies, and cultural history, which privileges the daily lives of "ordinary" people, we can trace an ideal model.

This "ideal model" would therefore be a model that would propitiate integration between these varied possibilities of approaches. As far as feasible, the interesting would be to consider the social and economic structures as the basis of analysis, since these are configured as the main structure on which the other approaches will articulate in an overlapping manner.

4. On the teaching of historical ethnobiology

Having before the eyes the necessity to work to integrate structure, culture, daily life, and microhistory, the teacher in historical ethnobiology has the chance to make a teaching practice more selective in relation to content selection.

Considering this perspective, teachers should have the attitude of avoiding a teaching practice that values exclusively political history, since it represents a part and not the totality of the history of human societies. Likewise, teachers should not have the attitude of belittling the historiography exercised in other times, privileging only the recent historiography. Dealing exclusively with the history of daily life detached from the study of economic structure and social institutions implies leaving students without reference. Thus, understanding a traditional approach constitutes a precondition for understanding other approaches that scape from this pattern. In the scope of these reflections, it is essential to discuss, even in a succinct way, traditional teaching and current tendencies in the practice of teachers who will deal with history, sources, and ethnobiology. It is important to think that education was defined at the beginning of the 20th century by Émile Durkheim as being an action of the adult generation, which would be the great diffuser of cultural traits on the young generations, who would be preparing to integrate themselves in their groups social (Durkheim, 2011). The purpose of education would be that the knowledge and values accepted by the mature stratum of society be apprehended by the younger strata, so it was expected that these would perpetuate the social life established over the past generations. At that time of history, in the educational sphere, students had no room to develop self-knowledge with independence, not even creativity.

This model of thinking about education, in general terms, aligns with traditional pedagogy. Going a little deeper into the history of pedagogical currents, the use of the expression "traditional pedagogy" refers in a more particular way to Johann Friedrich Herbart's theory (Hilgenheger, 2010). However, we can also situate it through other conceptions, such as those postulated by Durkheim and others, who conceive education as a one-sided practice.

This pedagogical current brings as main guidelines as follows: the organization of classroom space is so students stay turned to the teacher; the role of the school is to prepare the student for intellectual skills and moral behavior; the contents comprise the knowledge and social values that have been socially accumulated; the methods applied are verbal exposure and repetition of exercises by memorization; the teacher is considered the only one to play an active role in the teaching-learning process, which means in the relationship the teacher's authority predominates; and, finally, the assumptions seek to ensure learning is a receptive and mechanical performance.

Although it is delicate to situate an educational current in time, as there is a confluence of pedagogical conceptions in the classroom, some authors seek to place them in periods that are related to their maximum of acceptance in educational practices. The formulation of the traditional pedagogical current has been situated in the 19th century. Despite this dating, this teaching method seems to have been adopted already in the Middle Ages. With this long historical trajectory, this traditional liberal trend is also still very much alive in today's teaching environments.

Following the paths of this traditional approach, teaching and learning in historical ethnobiology address historical issues by making connections between facts, period of time, geographic determination (village, country, etc.) and names and memorizing this information for the learning evaluation process. To fully adopt this practice signifies being far from providing a space for understanding, reflection, critical analysis, comparison, and discussion. The basis of the teaching-learning process is based on the memorization of the content and simplistic and non-reflective answers.

Contemplating a discipline of ethnobiology or a specific discipline of historical ethnobiology, so the teaching of the content of historical ethnobiology does not follow this model in which the formal contents (the knowledge) must be decorated and forgotten immediately after the end of the content taught, it is necessary to create pedagogical strategies that make these formal contents significant to the student. Thus, what should be put into practice is a pedagogical action that encourages students to seek an understanding of the past facts and their multiple reliefs, establishing relations with their reality or with the present moment and, thus, achieving a deeper understanding society at its various levels, starting from itself to the global community. Some parameters can help us to think about the way of teaching a historical ethnobiology that goes in this direction, such as the following: "to use methods of research and production of texts of historical content learning to read different written registers, iconographic, etc.; recognize changes and permanences in the human experiences present in the reality of students or in other communities, that are near or distant in the temporal and spatial scale; to know and respect the way of life of different social groups, in different times and spaces, in their cultural, economic, political, and social manifestations, seeking to recognize similarities and differences between them; [and, finally,] to organize some historical and cultural repertoires that allow them to locate events in [...] time, to propitiate the formulation of explanations related to some issues of the present and the past" (Brasil, 2000).

These propositions of teaching practice can only be present in the teaching-learning process to the extent that the teacher assumes a conception of learning historical knowledge that goes beyond memorization and reproduction of the content. Following, it will be presented a breakdown of the work based on documentary sources.

5. Working with documentary sources

It is necessary to resume, albeit quickly, the conceptualization of source or historical documentation. Traditionally, the idea of a document with historical value exclusively covered official documents. More recently, there has been an expansion and different repertoires have moved to the category of historical document. In this way, an oral testimony, a photograph, a drawing, a recipe notebook, a school notebook, a personal or official letter, and so many other examples could integrate this list of what can be considered, often, as or more significant than an official written register, depending on the situation and the enunciative content of the documents.

Another progressive change has occurred in relation to the way documents are analyzed. Ethnobiologists, within a historical perspective, must often read between the lines, not only paying attention to what is written or represented, what is being said, but also perceiving what is not explicit in the discourse — what is unconscious or even has an occult meaning — such as judgements or a particular worldview.

In the content teaching practice that deals with history, the use of documents is essential. Even if the teacher chooses to work with students with a single documentary source on a determined subject, without exhausting all other sources, direct work on a unique document can provide students with discoveries of traces of the past through this approximation with the historical source.

Nevertheless, it is worth mentioning that contact with historical sources does not represent a direct interaction or access to the vestiges of social memory. The document stands as a interface between our present historical consciousness and events that are no longer directly accessible. In this way, what happens is that the contact and reading of documents result in the reconstruction of a particular sequence of the past through the cognitive structure of people. But this does not represent a process in which the reading and human interpretation distort the historical reality of the facts and that documents. What happens is that the documents answer the questions we ask them, and those questions are loaded with our current guidelines and concerns at the present time.

An interesting practice to be adopted in the classroom is to select documents that can be presented along with questions for the debate of students in a group or with the whole class. This practice will result in an immersion of students regarding the historical documentation, as well as providing the teacher with the construction of a didactic collection. In addition to historical documents, produced at a given time, texts produced later, such as those prepared by ethnobiologists or historians or even reports from current newspapers and magazines that address the theme of the past that were chosen to be discussed, can also be used. When making this presentation of a documental set to the students, it is necessary to take into account that the historiographic and journalistic analysis of a past event represents a document about a past era and is impregnated by the present historical consciousness about this past time. In this sense, it is essential that the teacher explain to the students about the origin of the material and the reason of analyzing it considering the pedagogical proposal of the class. To talk a little more about working with historical documentation in the classroom, we will cover some suggestions subsequently on how to use some types of sources as a teaching resource.

5.1 Type of source: Official documents

They are those documents that receive the approval of the State and are issued by public agencies or private entities. They include certificates, deeds, power of attorney, dispatches, laws, money notes, and others. These documents have long been considered the most reliable and important types of sources. This is due to the description of the facts in general with a presentation of information in an objective, clear, and specific way. However, even this typology of documents may contain subjective aspects or may even hide memory traces. It may occur that during the analysis the researcher comes to the conclusion that there is information relevant to the composition of the document that has been occulted. In view of a situation such as this, in which the absence of information in a specific document is observed, the researchers can reflect on the reason why the author of the documents chose to occult these elements. In this sense, it is also up to the researchers to formulate hypothesis about these issues because the document is leaving some important information without being expressed.

The use of this type of document in didatic practices is beneficial and an identity interesting to the teacher for the access we can have to several official documents. One last important reasoning to be raised would be that it is important to reflect with the students that the document is a representation; that is, it is not the possession of a document that provides someone an identity, considering for example the personal identification card. But on the contrary, since each person has an identity, it is that one is entitled to a document that attests to this fact.

5.2 Type of source: Letters

Two types of letters are usually identified by historians, so-called official letters and personal letters.

Within the ambit of official letters, those addressed to the authorities, such as religious or political are included, normally with informative content on issues of interest to the State. They are composed by elements of objective character, such as dates linked to events that occurred. But also, the elements of subjective character are present, as would be the case of reports on Brazilian nature, referencing the celestial idea. In this example, we have implied the mentality of the age of discovering something that is subjective to the speech written in letters from this historical period. When working this type of letter with students, it is important that during the analysis, students can identify the objective and subjective elements contained in the text, that is, which are the ideas placed in an explicit way and which are not being expressed by the letter's author. In this process, the effort is to go beyond what is evident in the textual composition and promote discussions among students, allowing time for each one to talk about what they can perceive through their reading and analysis.

Personal letters, on the other hand, are those written with personal interests and content geared to expressing issues that involve people's private lives. Considering the history of everyday life, these letters are fundamental to the approach of other aspects of human life. In this way, the analysis of the personal letters is interesting because they can put us in touch with the most intimate issues, the mentality, the values and other more personal reliefs of social actors who left their ideas registered in these type of document.

5.3 Type of source: Journalistic texts

In the process of historical research, for a long time, journalistic texts were taken as sources of minor relevance, for expressing a subjectivity. Normally the journalist does not have an objective writing, letting enter the text not exclusively what is occurring, but their own opinions, positions, and subjective values also integrate the journalistic text.

More recently, this type of text has been valued as a historical source. This is due to a broader understanding of how to make history. In the measure that the researcher of history does not seek to know only about the facts *per se*, feelings, meanings and personal values of other periods are also taken into account in the search for a better understanding of the past. In this type of text, the subjectivity present in the expression of information can be compensated by the possibility of access to a set of information that will be published in more than one editorial of the newspaper. Thus, the researcher can follow the evolution of the facts in a progressive way and can perceive the incorporation of new outlines given to the situation. When using journalistic texts in didactic interventions, we must keep in mind that it will be necessary for the teacher to enable students to understand the text from a broader historical context, which places the right period covered in the document. Along with this broader view, students should also be encouraged to do the reverse exercise, that is, to resort to a more reduced view, paying attention to the details of the facts and, thus, they can build a critical analysis of the document. It is the exercise of analyzing from the knowledge we have today about that past fact and returning to the journalistic text and discussing in minutiae the information that is being brought to us.

5.4 Type of source: Literary texts

More recently, literary texts have again been used as a historical source by researches. From the traditional perspective, the literary text served strictly as a certification of what had already been verified in the official documentation. There was then an additional character given to these documents, and this attitude constitutes the differentiated position today assumed about the value of literary texts. Like the literary text, official documents are also endowed with subjective elements, which can be studied through comparative analysis between different typologies of sources. In favor of the analysis of literary texts, we can say that they provide us information about several contexts, which help us to understand their nuances in a different way from the official document. Certain types of information do not have space in official documents, which can however be apparent in literary texts when they mark elements related to the daily life of cultural groups of a given time in the narrative. Thinking about teaching practice, the approach of this type of document in the classroom can provide reflection on the permanence and change of memory considering a temporal scale, by proposing to students a comparative analysis between the past and the present.

5.5 Type of source: Oral reports

Considering newer traces of memory for the study of historical ethnobiology represents a possibility to promote the integration between generations of a given society. Through the report of older people who have lived historical moments that we wish to research, we can hear from them what they can and wish to talk about their experiences of life. Through these reports, students can have the opportunity to contact a more vivid description of the past.

Although there is all this special condition of a more recent living memory about a particular fact of the past, care must be taken when dealing with it. When starting the activity, the teacher should talk to the students about the fact that human memory is selective, and that, that, therefore, it is not uncommon for the person not to bring to the spoken discourse all the details about the facts. In this process of telling about the past through spoken memory, what we are told by the people selected to tell us about a fact is told through many selective layers, represented by our personal values, worldviews, beliefs and other memories that are associated to the fact that is being reported.

In the classroom, the teacher who chooses to work with oral history should give the students a plan to be used during the application of interviews to direct the activity. After this stage, the teacher should promote a debate in which a comparison is made between the experiences described by the interviewees and other documentary sources that deal with the same subject. Through oral testimony the fact gains dynamism, comes to life. However, the use of other sources creates the possibility of achieving a criticism of memory that takes us to a depth in the analysis of the facts. So, this dynamic will allow a practical experience of students in relation to the integration and complementarity between information from different sources.

5.6 Type of source: Iconographic source

Images allow us to look with the eyes of the past, that is, through the gaze of people from past societies. They give us access to scenes, environments, social actors, nature, landscapes, etc., from past times. They are period images that reveal a multiplicity of spaces, people and times.

The didactic work with images should be done with a careful approach. Some aspects are involved in the use of images. The first aspect is that the images are not void of meaning in their message. This means that they reveal the artist's look, or in the case of photography, of those who took it. In the image is expressed what the artist or layperson wants to call our attention to. Thus, as a basic procedure of analysis, we must be aware of what is hidden or what is being diverted in the image. From this attitude of critical analysis of images, we can bring to the fore the discussion of aspects of the larger context of that representation and give visibility to themes and issues that have become invisible or that have been forgotten by history.

Another aspect of which one should be attentive when using images is to avoid an anachronistic conception of the image. Some situations fit this situation at the time of analysis.

One case would be, for example, of an author of an oil on canvas who painted a scene in which the artist himself did not experience and whose fact portrayed happened in the time before that of his work. In this case depicted, the image has a double reference: the past event (fact of the past), which represents the theme of the painting, and the historical and artistic consciousness of the period in which the artist was (or is) situated.

Another procedure when analyzing images would be the identification that an artist painted scenes of his own time. In this case, there would be a single order of references since the theme of the work and the historical consciousness of the artist coincide in terms of the time of representation and production.

It is also worth mentioning that the use of moving images is also an interesting possibility. However, one should evaluate the length of films and documentaries. Special attention should be paid to this issue so it is feasible to combine the display with the practice of immersion exercises and debate among students. Observations on image analysis also apply to moving images. This means that there are distinct values between a film that portrays a natural theme, typical of the time in which it is filmed, and another in which the filmmaker reproduces a scene from the past, where scenarios and actors are used to compose an era.

6. Evaluation practices and methods in the teaching of historical ethnobiology

Although there are several discussions in the theoretical plan about the importance of learning assessment, many teachers perpetuate formal evaluation models. This procedure seems to reveal that there is a need to analyze both the students' level of learning during the formal teaching process and the teacher's pedagogical practice. It may indicate that the reflections about the evaluation method are focused on understanding the evaluative action, and that it is far beyond the simple conference through the application of tests.

The evaluation, then, aims to judge the extent to which educational objectives have been achieved. In this sense, there is an approximation between the application of the terms "appreciation", in the conceptualization of "evaluation" presented by Libâneo (1994), and "evaluation", in the definition of Haydt (2006). That means that evaluating for these educators is not applying a test or assigning a grade, but it means thinking about the students' cognitive and affective development and their potential abilities.

In this conception, an evaluation can be carried out before, during, or after the teaching process, thus providing a reflection on the progress of student learning in different pedagogical moments. Depending on the time when the evaluation occurs, it receives a different denomination from the others. Thus, the diagnostic evaluation analyzes whether the student already knows; formative evaluation takes place during the teaching process and intends to create a possibility of methodological adjustment during the teaching-learning process; and the control evaluation analyzes the entire process to perceive if the educational objective was achieved.

In the understanding of a formative and non-summative assessment, the educational objective is not to assign grades, but to focus on the reasoning developed during the resolution of situations proposed by the teacher. What matters for this type of assessment is the understanding of the mental schemes that students develop to build their own knowledge.

To achieve an evaluation that does not consider only a reproductive learning with memorized and unreflected responses, we need to understand the evaluation as a process. This means that the teacher must realize that all the activities developed during the course are related to some form of evaluation. The tests, the written or oral exercises, that is, formal evaluations do not configure a pause in the educational-evaluation process. These evaluations represent different stages in the construction of knowledge. The process of reworking the contents as well as learning how to express them, however, can be enriched by other evaluation methodologies. This situation can be commented taking as an example what can happen during an expository class. In this situation, the teacher can, at the same time, present the content, observe the attention, interest and participation of the students and also evaluate them. In group activities, students can be permanently evaluated through their self-assessment, the evaluation of the other members of the group, and, finally, by evaluating the results of the tasks they are performing. The formative evaluations may also include issues presented to the students such as text interpretation, schemes, summaries, self-assessment, interviews, and observations, among others.

A main model for data collection is the observation that the teacher makes of each student. In this case, the teacher builds an evaluation form in which the observations about the behavior manifested by the students and also about their participation in the proposed activities are plotted. Another concern in this formative process is the care to provide feedback to the student about the outcome of this evaluation process.

What is important to reflect on the most appropriate form of evaluation for each class and content is that formal evaluations are the result of an evaluative process, not an instant of evaluation. With this predisposition, teachers should seek to expand their possibilities of evaluation methods with their students. Thus, written or oral tests or exercises should be combined with other techniques, such as group dynamics, rounds of discussions, production of schemes or concept maps, etc. In this less traditional procedure of evaluation of the students, it is appropriate in the tests or in other types of written evaluations to favor the open issues. These formulations guarantee creative space on the part of students, not requiring the same answers based on a practice of memorization. In the teaching of historical ethnobiology, the fundamental thing is that students learn to think historically.

7. Activity suggestions for teaching historical ethnobiology

Next, we will present some suggestions for practical activities with sources to serve as ideas for teaching practice in the approach of historical ethnobiology. Certainly, the teachers will know the best adequation to their reality and will be aware that what we bring here is not an exhaustion of the possibilities of working the theme with their students. Using their experience and creativity, the teachers will be able to refine or develop other proposals that will integrate the set of teaching techniques used in their historical ethnobiology classes.

The important thing in this process of planning and applying the teaching technique is to think that we must create conditions for our students not only to decorate, but that they are encouraged to think historically. The ambience in this case is fundamental. And in this sense, the teacher should seek to have the role of mediator, and thus work on creating an educational environment in which students feel they are participating in real historical research. The essential thing is that students develop the ability to analyze and interpret the meaning of the contents present in a given historical source.

The activities developed with documents consist of concrete examples of typologies of historical sources and debates that can be generated from the work with them.

Thus, as examples of activities with documents, we can suggest the comparison of images, the analysis of letters and chronicles, journalistic texts, beyond the work with historiographic texts, transcriptions of oral reports, among other possibilities.

Let us look in more detail in at least three initial suggestions for activities that will make room for teachers to create their own set of activities and teaching resources

7.1 Practical activity: Comparative study of images

As the view of historical ethnobiology is focused on the study of human societies over time and their relations with the elements of nature, an essential aspect for research is to consider the perception of changes and permanences of the memory traces.

Considering the teaching practice, the use of images as a didactic resource can be interesting to work on this perception. The observation and comparison of at least one duo of images, one image being older than the other, can provide students with a favorable condition for identifying changes and permanences. It is an activity to point out what has changed and what has not changed in the landscape, in people, or in what is being represented in the set of images presented to students.

Following, we will present an example of the use of images. The planning of this activity aims to lead students to think about changes in the appearance of a social actor of a specific cultural group.

- Activity description:

This didactic activity consists of three stages, which will be commented on next.

- First stage:

The teacher should take to the classroom a duo of images with elements that can evidence changes and permanences over time. The images should be sharp and should be shown to all learners. An observation time should be given to the students, having this action an invitation from the teacher asking them to compare the two images. After the observation period, the teacher instructs students with information that shows details about the images chosen for this stage of the activity. These comments can be, for example, these images are of the same person, they are photographs, or one was taken 50 years after the other; then, notice how the appearance of the person being represented has changed in that time interval.

- Second stage:

After this observation stage, invite students to check at home if they have old and current photos that represent elements that they want to raise to discussion in the classroom, always thinking about the concepts of change and permanence. It is worth reminding students that the images they will bring should be treated with the utmost care, as they are remembrances of the past.

In this phase of individual imagery research, students should collect the images and information related to them, such as who is being represented in the images, their dating, techniques, and how today the elements are represented in these images.

- Third stage:

In a second class, students should bring the images selected by them, and the teacher should then initiate a circle of conversations. Each student will be invited to present the information collected about the images selected for this analysis. The focus of the discussions should be the change in these images brought by time.

7.2 Practical activity: Analytical study of letters and chronicles

For the teaching of historical ethnobiology, the use of letters and chronicles as didactic material is interesting because they bring information written by authors who narrate events from the same period in which they lived.

- Activity description:

  For the work with this typology of source, a complete analysis can be made, or the execution of the activity can be simplified through the analysis of few sentences. The activity with letters and chronicles encompasses three stages, which will be described next.
- First stage:

  The teacher should select a theme to be addressed and then choose letters or chronicles that deal with what is intended to be addressed in the classroom.
- Second stage:

  Already in the classroom, the teacher should make a previous explanation about the documents that will be read by students, giving a succinct description about them. Next, learners will have time to read the selected documents, asking them to highlight the textual elements that evidence the theme that is being focused.
- Third stage:

  For the closing of the activity, a conversation should be proposed in which students will expose about what characterizes each document in terms of historical characters, environments, mentalities, values, in addition to other important nuances present in the expression of ideas contained in the narratives read. On the technical side are details related to the authorship, dates, linguistic style, calligraphy, and technical characteristics of the document.

7.3 Practical activity: Study with newspaper text

For the teaching of historical ethnobiology, journalistic texts can be interesting sources for working in the classroom. A challenge, however, will be access to these texts, depending on the access that we have to files, libraries, or even the internet depending on the local reality. But we make room here for this type of source since it is configured as a very rich material for the research and teaching of historical ethnobiology.

- Activity description:

  This activity will aim to plan the development of in-depth reading skills, identification of textual elements, transcription, and composition of sources. The activity can have three stages, which are described next.
- First stage:

  The starting point will be the choice of a theme that we want to discuss. A second step will be the selection of one or more journalistic texts that dialogue with each other or that bring nuances of the selected themes for discussion with students.

- Second stage:

  In the classroom, the teacher will present the set of selected texts to the students, and then ask them to do an in-depth reading of them. At this point, students should also identify the elements present in the text that refer to the selected theme. After the identification, the elements must be transcribed on a separate sheet of paper. For the transcribed excerpts, students must register the composition of the complete source citation. Thus, a personal notebook will be created with these excerpts and sources.

- Third stage:

  At the end of this research phase, the teacher can generate a conversation circle in which students expose their experiences about the research process experienced, commenting on their skills and difficulties, seeing the correct forms of transcription and composition of sources, and another discussion should also be made on the theme expressed through the excerpts transcribed by the students.

8. Conclusive considerations

At present, we can think of a teaching of ethnobiology, especially of historical ethnobiology, that is guided by the exercise of a critical position in its role of educator. Within this bias, the conception of teaching in ethnobiology should encourage the existence of an educational process that wishes to promote and value the autonomy with respect through a pedagogical action in which the teacher assumes the role of mediator with the students.

Our discussions revolved around this critical reflection on teaching and on the constitution of materials and practical activities. Some teaching practices have been suggested without, however, being configured as a single direction to be followed. On the contrary, what we hope is that teachers will accept these considerations as an invitation to creation and criticism in the construction of personalized portfolios adapted to the most diverse realities of teaching.

References

Agostinho, S., 1984. Confissões, 3rd. Abril Cultural, São Paulo.
Brasil, 2000. Parâmetros Curriculares Nacionais. Ciências Humanas e suas tecnologias. MEC, Brasília.
Durkheim, É., 2011. Educação e Sociologia, 3rd. Editora Vozes, Petrópolis.
Haydt, R.C.C., 2006. Curso de didática geral. São Paulo, Ática.

Herodotus, 2014. Histories. Viking, New York.
Hilgenheger, N., 2010. Johann Herbart. Fundação Joaquim Nabuco, Editora Massangana, Recife.
Libâneo, J.C., 1994. Didática. São Paulo, Cortez.
Megill, A., 2007. Historical Knowledge, Historical Error. Chicago University Press, Chicago.
Tucídides, 2001. História da Guerra do Peloponeso, 4th. Editora Universidade de Brasília, Brasília.
Voltaire, 1769. Historia de Carlos XII, Rei de Suecia. José de Aquino Bulhoens, Lisboa.
Von Ranke, L., 2011. The theory and practice of history. Routledge, London.

CHAPTER NINE

Final considerations

Contents

1. A possible theoretical-methodological path: closing the discussion

In the midst of the reflections that until now have been drawn, we could go to the closing of the book with the idea of an identity search that takes place through ethnobiologic study. This research seeks to understand the relations between people and elements of nature over time through a patrimonial diversification existing today that arrives at us with overlapping layers of information and meanings. It is an incessant search for the preservation of cultural patrimony and social memory.

We have observed through previous chapters that the objects are testimonies, therefore, are evidences; they are signs and have meanings. To the object that represents the material culture (document of various types of support), multiple layers of information are aggregated since this constitutes a mediator of the most different social aspects. Individuals and societies, their singularities and plurality are represented in the objects. The layers they bring in themselves are well particularized. They can have the meaning of the constituent materiality of revealing and understanding the technique, describing the form, and identifying the function. Thus, the object that is source of study in historical ethnobiology is at the same time ethnography, history, chemistry, physics, mathematics, technique, and art.

In the research path starting from the documentary sources the way is not at all simple for the evocation of information. Thus, we have to develop our research seeing the source beyond its materiality, paying attention to every information that the object in study can give us in return.

In the methodological course, the researcher is expected to have an accurate technical procedure through which it allows deep dimensions (relations and meanings) of the sources to be disclosed by the researchers.

Historical Ethnobiology
ISBN: 978-0-12-816245-3
https://doi.org/10.1016/B978-0-12-816245-3.00009-3

In the guiding principle, we must make the effort to construct a set of information as complete as possible, which we can observe in the study model proprosed by Susan Pearce (2005). In this model, she calls attention to several aspects related to objects, which we extend to any source that we belive is also applicable to this scale, which includes the material plane of object (source), the design, the characterization, the history, the micro and macro context, localization, meaning, and interpretation of the role of the object (source) in the social plot. Logically, these approaches are not absolute and cannot be solved by all sources in which we are researching. However, to have the conscience about the importance of working on such approaches is fundamental for us to have an in-depth research.

In this research process, all traits, principally the peculiar, have to be considered and understood as guiding elements of a social content. Thus, to do the reading of the object under analysis, it is necessary to be thorough, to dedicate time in identifying all the details that will give us clues to the real meaning of that object within a culture. In this way, we would evaluate whether it would be an object that could be a testimony of social memory, which would deserve to be part of a biocultural collection.

At this point, it should be commented that the importance of the collection is no longer centered only on the documentary source itself, but also on its representation constructed by an intelective process using language as a means of expression and transmission of knowledge of a specific object. It is a semantic interpretative procedure that involves producing, systematizing and treating information through phonetic signals, that is, language. Through language, it becomes possible for the researcher to express the full range of social memory through the meaning relations of objects with their traces of permanence or discontinuity in a given culture. Thus, in this process of transforming an object into a document, we come into contact with layers of information present in the documentary sources, which can bring to us the most diverse types of interpretations. There is a myriad of sources and the set of information associated with them is of historical origin, as they represent knowledge and social memories from a more distant or recent past. Through this treatment of information and contextualization of the supports of social memory, the documentary sources are dynamized in a new context, which is the space of the institutions of memory. By this research movement, the formation of biocultural collections favors society to receive preserved material and immaterial cultural knowledge from the past, as well as current information.

What is worth emphasizing is that objects go far beyond our existence. The sources remain, but their memory does not always remain, because the social memory related to the objects (sources), which are protected in the institutions, can be lost as we seek traces of memory that we intend to evoke. The number of objects, texts and ideas that can be recovered has no limit and casual forgetfulness becomes part of the remembrance process.

To conclude the reflections made throughout the chapters of this book, we can conclude that the construction of the scientific narrative is essential for research in historical ethnobiology, as it has the role of allowing the transmission of past or recent information from social memory about relations of peoples with nature, which are full of meanings. Criticism of memory sources is the principle that ethnobiologists must consider during narrative production. Considering the meanings emerging from this criticism of the sources, we can ask ourselves: How do we find value in objects (sources)? How can we give meanings to them in our scientific narratives? How could social memory really be contemplated in these sources? It may be that only with a critical analysis of information it would be possible to construct a narrative of social memory and, thus, exhibit it in biocultural collections.

The objects safeguarded in memory locations are documents of changes in social memory and, at the same time, of the permanence of cultural traces and are connected to a narrative of the profound structure of collective memory. Therefore, the elements that compound the collections represent a documentary set, a memory about the forms of knowledge, about historical evidence and also respond to concerns. And objects, when placed in relation to each other, constitute documents to be processed by ethnobiologists.

In the perspective of historical ethnobiology, it is clear that this field of research meets the principle of complexity rather than logical simplification. We do not have the simplification of a methodological conduct, as it is the examination of the techniques of different disciplinary fields that impels us to make reflections and to seek to comprehend the concepts and instruments from which the disciplines are utilized to obtain their results. That way, it will be the specificity of each study that will require of the researcher a methodological outline based on this common base of procedures that we saw in the development of the ideas presented.

The theoretical-conceptual comments are not intended to fulfill the function of eliminating interfaces of thoughts between discourses of different areas of knowledge, isolating them into immutable categories that will serve historical ethnobiology. The certainty of methodological problems is more productive than walking a definitive path. The correlation of general concepts with theoretical and methodological instruments had the function of constructing a cognitive ordering, of constructing a panorama, in which document and information are intertwined as objects of study. In this sense, it is possible that the intersections and relations between concepts, theoretical and methodological instruments, and discourses can be maintained, at least at the operational level, more uniformly in different researches.

Reference

Pearce, S.M., 2005. Pensando sobre os objetos. In: Granato, M., Santos, C.P. (Eds.), Museus Instituição de Pesquisa, pp. 11–22. Mast Colloquia 7, Rio de Janeiro.

CHAPTER TEN

Suggested bibliography

Contents

This chapter is dedicated to present a suggested bibliography for further studies in the areas that compose the research in historical ethnobiology. Our aim is not to provide a complete bibliography, but to suggest how to start a bibliography, where the reader can find repertoires used as a source during the experiences of research that we already had as well as during the preparation of this book. Therefore, there will be bibliographic indications that we believe will be good to begin the immersion in the theme to which the structural design of this book is proposed.

Suggestions to additional readings about the history and development of ethnobiology

Albuquerque, U.P., Alves, A.G.C., 2016. What is ethnobiology? In: de Albuquerque, U.P., Nóbrega, R.R.A., Org. (Eds.), Introduction to Ethnobiology. Springer, Switzerland, pp. 3–7.

Alves, A.G.C., Albuquerque, U.P., 2010. "Ethno what?" – terminological problems in ethnoscience with special emphasis on the Brazilian context. In: Albuquerque, U.P., Hanazaki, N. (Eds.), Recent Development and Case Studies in Ethnobotany. Núcleo de Publicações em Ecologia e Etnobotânica Aplicada, Recife, pp. 67–80.

Anderson, E.N., 2011. Ethnobiology: overview of a growing field. In: Anderson, E.N., Pearsall, D., Hunn, E., Turner, N. (Eds.), Ethnobiology. Wiley-Blackwell, New Jersey, pp. 1–14.

Baker, H.G., 1970. Plants and Civilization, 2. ed. Wadsworth Publishing Company, Belmont.

Balick, M.J., Cox, P.A., 1996. Plants, People and Culture: The Science of Ethnobotany. Scientific American Library, New York.

Batalha, L., 1998. Emics/Etics Revisitando: "Nativo" e "antropólogo" lutam pela última palavra. Etnográfica 2 (2), 319–343.

Castetter, E.F., 1944. The domain of ethnobiology. Am. Nat. 78, 158–170.

Clément, D., 1998. The historical foundations of Ethnobiology (1860–1899). J. Ethnobiol. 18 (2), 161–187.

Cotton, C.M., 1996. Ethnobotany: Principles and Applications. John Wiley & Sons, Chichester.

Davis, E.W., 1995. Ethnobotany: an old practice, a new discipline. In: Shultes, R.E., Reis, S. von (Eds.), Ethnobotany: Evolution of a Discipline. Dioscorides Press, Portland.

Historical Ethnobiology
ISBN: 978-0-12-816245-3
https://doi.org/10.1016/B978-0-12-816245-3.00010-X

Ford, R.I., 1978. Ethnobotany: historical diversity and synthesis. In: Ford, R.I., Brown, M.F., Hodge, M., Merrill, W.L. (Eds.), The Nature and Status of Ethnobotany. Museum of Anthropology, University of Michigan, Ann Arbor, pp. 33–49.
Heiser, C.B., 1985. Of Plants and People. University of Oklahoma Press, Norman.
Hunn, E., 2007. Ethnobiology in four phases. J. Ethnobiol. 27 (1), 1–10.
ISE-International Society of Ethnobiology, 2014. Available from: <ethnobiology.net/>.
Lewington, A., 1990. Plants for People. The Natural History Museum, London.
Martin, G.J., 1995. Ethnobotany: A Methods Manual. Chapman & Hall, London.
McClatchey, W.C., 2005. Exorcising misleading terms from ethnobotany. Ethnobot. Res. Appl. 3, 1–4.
Posey, D.A., 1987a. Etnobiologia: Teoria e prática. In: Ribeiro, B. (Ed.), Suma Etnológica Brasileira – 1 Etnobiologia. Vozes/Finep, Petrópolis, pp. 15–251.
Posey, D.A., 1987b. Etnobioolgia y Ciência "Folk": Su importância para la Amazônia. Hombre y Ambiente 1, 7–26.
Wolverton, S., 2013. Ethnobiology 5: interdisciplinarity in an era of rapid environmental change. Ethnobiol. Lett. 4, 21–25.
Xolocotzi, E.H., 1983. El concepto de etnobotánica. In: Barrera, A. (Ed.), La etnobotánica: três puntos de vista y uma perspectiva. Instituto Nacional de Investigaciones sobre Recursos Bióticos, Xalapa, pp. 13–18.

About concept and principles aspects of historical ethnobiology

Albuquerque, U.P., 1997. Etnobotânica: uma aproximação teórica e epistemológica. Revista Brasileira de Farmácia 78 (3), 60–64.
Hastorf, C.A., Popper, V.S., 1988. Current Paleoethnobotany – Analytical Methods and Cultural Interpretations of Archeological Plant Remains. The University of Chicago Press, Chicago/London.
Medeiros, M.F.T., 2007. Fontes documentais do século XIX: fundamentos para a pesquisa etnobotânica hodierna. In: Barbosa, L.M., Santos Júnior, N.A. (Eds.), A botânica no Brasil: pesquisa, ensino e políticas públicas ambientais. Sociedade Botânica do Brasil, São Paulo, pp. 565–568.
Medeiros, M.F.T., 2009. Etnobotânica histórica: princípios e procedimentos. Núcleo de Publicações em Ecologia e Etnobotânica Aplicada, Recife.
Medeiros, M.F.T., 2010. Historical ethnobotany: an approach through historical documents and their implications nowadays. In: de Albuquerque, U.P., Hanazaki, N. (Eds.), Recent Developments and Case Studies in Ethnobotany. Núcleo de Publicações em Ecologia e Etnobotânica Aplicada, Recife, pp. 127–142.
Medeiros, M.F.T., Org., 2010. Aspectos Históricos na Pesquisa etnobiológica. Núcleo de Publicações em Ecologia e Etnobotânica Aplicada, Recife.
Medeiros, M.F.T., Albuquerque, U.P., Orgs., 2012. Dicionário brasileiro de etnobiologia e etnoecologia. Núcleo de Publicações em Ecologia e Etnobotânica Aplicada, Recife.
Medeiros, M.F.T., 2016. Historical ethnobiology. In: Albuquerque, U.P., de Nóbrega, R.R.A., Org. (Eds.), Introduction to Ethnobiology. Springer, Switzerland, pp. 19–24.
Salick, J., Konchar, K., Nesbitt, M. (Eds.), 2014. Curating Biocultural Collections – A Handbook. Royal Botanic Gardens, Kew.
Schultes, R.E., von Reis, S., 1995. Historical ethnobotany. In: Schultes, R.E., von Reis, S. (Eds.), Ethnobotany: Evolution of a Discipline. Dioscorides Press, Oregon, pp. 89–91.
Senna-Valle, L. de, Medeiros, M.F.T., 2014. Etnobotânica Histórica no Brasil. In: Silva, T.R.S., Moura, C.W.N., Lima, L.C.L., Santos, F.A.R., Org. (Eds.), Botânica na América Latina: Conhecimento, interação e difusão. Sociedade Botânica do Brasil, Salvador, pp. 310–316.

Silva, T.C. da, Medeiros, P.M., Balcazár, A.L., Araújo, T.A.S., Pirondo, A., Medeiros, M.F.T., 2014. Historical ethnobotany: an overview of selected studies. Ethnobiol. Conserv. 3, 4. https://doi.org/10.15451/ec2014-6-3.4-1-12.

About individual memory, social memory, and material culture

Abreu, R., 2008. Patrimônios etnográficos e museus: uma visão antropológica. In: Abreu, R. (Ed.), E o patrimônio? PPG em Memória Social da UNIRIO, Rio de Janeiro, pp. 33–57.

Breffe, A.C.F., 2005. O Museu Paulista: Affonso Taunay e a memória nacional. Fundação Editora da UNESP, São Paulo.

Belting, H., 2004. Pour une anthropologie des images. Gallimard, Paris.

Candau, J., 2005. Anthropologie de la mémoire. Armans Colin, Paris.

Dodebei, V., 2005. Memória, circunstância e movimento. In: Dodebei, V. (Ed.), O que é Memória Social? PPG em Memória Social da UFRJ, Rio de Janeiro, pp. 43–54.

Foucault, M., 1999. A ordem do discurso, fifth ed. Loyola, São Paulo.

Granato, M., Santos, C.P. Museus Instituição de Pesquisa. Mast Colloquia 7, Rio de Janeiro.

Namer, G., 1987. Instituitions de memoire culturelle. In: Namer, G. (Ed.), Memoire et societé. Méridien, Paris.

Rossi, P., 2004. A chave universal: Artes da memorização e lógica combinatória desde Lúlio até Leibniz. Edusc, Bauru.

Yates, F.A., 1993. L'Arte Dela Memoria. Einaudi, Torino.

About human communities, nature, and culture

Alves, A.G.C., Souto, F.J.B., Peroni, N., Org., 2010. Etnoecologia em perspectiva: Natureza, cultura e conservação. Núcleo de Publicações em Ecologia e Etnobotânica Aplicada, Recife.

Anderson, B., 1991. Imagined Communities. Verso, Londres.

Baker, H.G., 1968. Las plantas y la civilization. Herrero Hermanos Sucesores S.A., Ciudad de México.

Bernardez, G., 1981. Ecologia y paisaje. Hermann Blume, Madrid.

Bourguinon, E., 1979. Psychological Anthropology: An Introduction to Human nature and Cultural Difference. Holt, Rinehart and Winston, New York.

Eriksen, T.H., 1993. Ethnicity and Nationalism: Anthropological Perspectives. Pluto Press, Londres.

Freilich, M., Org., 1972. The Meaning of Culture. Rand Xerox Publishing, Lexington.

Geertz, C., 1973. The Interpretation of Cultures. Basic Books, New York.

Klanovicz, J., 2002. Uma etnobotânica no México. Episteme 15, 133–136.

Goodenough, W.H., 1971. Culture, Language and Society – McCaleb Module in Anthropology. Addison-Wesley Publishing Company, Reading.

Johns, T., 1990. With Bitter Herbs They Shall Eat it: Chemical Ecology and the Origins of Human Diet and Medicine. University of Arizona Press, Tucson.

Le Vine, R.A., 1973. Culture, Behavior and Personality. Aldine Publishing, Chicago.

Le Vine, R.A., Org., 1974. Culture and Personality: Contemporary Readings. Aldine Publishing, Chicago.

Lifschitz, J.A., 2011. Comunidades tradicionais e neocomunidades. Contra Capa, Rio de Janeiro.

Solinger, D., 1975. Aspects of Language, 2. ed. Harcourt Brace Jovanovich, New York.

Wagner, R., 1975. The Invention of Culture. Prentice-Hall, Englewood Cliffs.

About historical sources and documental analysis

Alves, R.R.N., Medeiros, M.F.T., Albuquerque, U.P., Rosa, I.L., 2012. From past to present: medicinal animals in a historical perspective. In: Alves, R.R.N., Rosa, I.L., Org. (Eds.), Animals in Traditional Folk Medicine. Springer-Verlag, Berlin, pp. 11–22.

Chabot, V., 1992. Des nomes pour la description des documents d'archives: la normalization en archivistique. Québec: Documentor.

Couture, C., Rousseau, J.-Y., 1998. Os fundamentos da disciplina arquivística. Publicações Dom Quixote, Lisboa.

Kobashi, N.Y. (Ed.), 1989. Análise documentária: considerações teóricas e experimentações. FEBAP, São Paulo.

Kobashi, N.Y., 1996. Análise documentária e representação da informação. Informare 2 (2), 5–27.

Lodolini, E., 1993. Archivística: principios y problemas. ANABAD, Madrid.

Medeiros, M.F.T., 2014. Procedures for documentary analysis in the establishment of ethnobiological information. In: Albuquerque, U.P., Cruz da Cunha, L.V.F., Lucena, R.F.P., Alves, R.R.N., Org (Eds.), Methods and Techniques in Ethnobiology and Ethnoecology. Springer, Switzerland, pp. 75–86.

Medeiros, M.F.T., Alves, R.R.N., 2017. Studying ethnozoology in historical documents. In: Alves, R.R.N., de Albuquerque, U.P., Org. (Eds.), Ethnozoology: Animals in Our Lives. Elsevier, pp. 45–52.

Morin, E., 1977. La method: la nature de la nature. Édicions du Seuil, Paris.

Pescador Del Hoyo, M.C., 1993. El archivo: instrumentos de trabajo. Ediciones Norma, Madrid.

Schellenberg, T.R., 1974. Arquivos modernos: princípios e técnicas. Editora da Fundação Getúlio Vargas, Rio de Janeiro.

Index

Note: 'Page numbers followed by "*f*" indicate figures and "*t*" indicate tables.'

S

T

U

Made in the USA
Monee, IL
13 September 2026

62977007R00120